ALSO BY MARGALIT FOX

Talking Hands:
What Sign Language Reveals About the Mind

THE RIDDLE OF THE LABYRINTH

THE QUEST TO CRACK
AN ANCIENT CODE

MARGALIT FOX

ecco
An Imprint of HarperCollins*Publishers*

HarperCollins books may be purchased for educational, business, or sales promotional use. For information please write: Special Markets Department, HarperCollins Publishers, 10 East 53rd Street, New York, NY 10022.

FIRST EDITION

Designed by Suet Yee Chong
Endpaper image by cybervelvet/Shutterstock.com

Library of Congress Cataloging-in-Publication Data has been applied for.

ISBN 978-0-06-222883-3

13 14 15 16 17 OV/RRD 10 9 8 7 6 5 4 3 2 1

The mental features discoursed of as the analytical are, in themselves, but little susceptible of analysis. We appreciate them only in their effects. We know of them, among other things, that they are always to their possessor, when inordinately possessed, a source of the liveliest enjoyment. As the strong man exults in his physical ability, delighting in such exercises as call his muscles into action, so glories the analyst in that moral activity which disentangles. . . . *He is fond of enigmas, of conundrums, of hieroglyphics; exhibiting in his solutions of each a degree of* acumen *which appears to the ordinary apprehension praeternatural. His results, brought about by the very soul and essence of method, have, in truth, the whole air of intuition.*

—EDGAR ALLAN POE,
"THE MURDERS IN THE RUE MORGUE," 1841

CONTENTS

BOOK THREE

THE ARCHITECT

LIST OF ILLUSTRATIONS

INTRODUCTION

THIS IS THE TRUE STORY of one of the most mesmerizing riddles in Western history and, in particular, of the unsung American woman who would very likely have solved it had she only lived a little longer. It is the account of the half-century-long attempt to decipher an unknown script from the Aegean Bronze Age, whose bare-bones name, Linear B, belies both its bewitching beauty and its inexorable pull.

I first encountered the tale of Linear B more than thirty years ago, as a moony adolescent, and it has lost none of its mystery or narrative power since. At its center was a set of tablets, buried for almost three thousand years and first unearthed only at the dawn of the twentieth century. Dating from the second millennium B.C., the tablets were inscribed with a set of prehistoric symbols like no writing ever seen. Despite the efforts of investigators around the globe, no one could discover what language they recorded, much less what the curious inscriptions said.

The decipherment of Linear B came to be considered one of the most formidable puzzles of all time, and for five decades, some of the world's most distinguished scholars attempted without success to crack the code. Then, in 1952, the tablets were deciphered seemingly in a single stroke—not by a scholar but by an impassioned amateur, an English architect named Michael Ventris. Young, dashing, and brilliant, Ventris had a long obsession with the tablets, a prodigy's gift for languages, and, as I would later learn, a sorrowful history. His life, to my teenage self, was the stuff of high romance. Still more romantic was the fact that his great triumph culminated in tragedy: In 1956, just four years after solving the riddle, Ventris died at the age of thirty-four, under circumstances that remain the subject of speculation even now.

But as captivating as it was, the story I knew—the only story anyone knew—was incomplete. A major actor in the drama was missing: an American woman named Alice Elizabeth Kober. Working quietly and meticulously from her home in Brooklyn, Kober was by the mid-twentieth century the world's leading expert on Linear B. Though largely forgotten today, she came within a hair's breadth of deciphering the script before her own untimely death in 1950.

Alice Kober's story is presented here in full for the first time. As her published papers and private correspondence make plain, it was she who built the foundation on which Ventris's successful decipherment stood, and it is clear that without her work Linear B would never have been deciphered when it was, if at all. In recent years, Kober's role in the decipherment has been likened to that of Rosalind Franklin, the English scientist now considered the unsung heroine of one of the most signal intellectual feats of the modern age, the mapping of the molecular structure of DNA by Francis Crick and James Watson.

What makes Kober's achievement especially striking is that she did her groundbreaking work entirely by hand—sitting night after night at her dining table with little more than paper and ink—without the aid of "IBM machines," as she dismissively called them. Yet for several reasons, not least among them that history is nearly always written by the victors, her contribution to the unraveling of Linear B has remained almost completely absent from the historical record.

Until now, only two slender histories of the decipherment have been published, *The Decipherment of Linear B* (Cambridge University Press, 1958), by John Chadwick, and *The Man Who Deciphered Linear B: The Story of Michael Ventris* (Thames & Hudson, 2002), by Andrew Robinson. Both of these lovely books are known primarily in Britain, and both—Chadwick's especially—devote comparatively little space to Kober. They could hardly have done otherwise: Kober's private writings, including her decade-long correspondence with other Linear B scholars, as well as her own painstaking analysis of the script, thousands of pages of documents in all, became available only recently. As a result, thanks to the newly opened archive of her papers at the University of Texas, this book can offer the first complete account of the decipherment. It is not meant to supplant either Chadwick's book or Robinson's, to both of which I am deeply indebted. Rather, it is meant to complement them, fleshing out the little-known American contribution to this captivating international puzzle.

"I DON'T LIKE the idea of getting paid" for scholarly writing, Kober said in 1948. "If I wanted to make money writing, I'd write detective stories." That, as it turns out, is precisely what she *was* writing: Read today, her work is a forensic playbook for archaeological decipherment. *The Riddle of the Labyrinth,* which centers on the cryptanalytic process involved in unraveling an unknown script, is a paleographic procedural, following the work of Kober and others step by step as they solve a riddle that had defied solution for more than half a century.

This book is also an amplification—even a refutation—of the few, brief biographical sketches of Kober that have appeared in published accounts of the decipherment over the years. Because the writers had none of her personal correspondence on which to draw, they were obliged to conjure Kober whole from her few, rigorous published articles. As a result, these sketches inevitably leave the reader with the impression of a stern, humorless woman who had little passion for anything outside the serious enterprise of deciphering Linear B.

"In the words of Ventris written after the decipherment, her approach was 'prim but necessary,' " Andrew Robinson has written in *Lost Languages: The Enigma of the World's Undeciphered Scripts,* published in 2002. "To go further would require a mind like his that combined her perseverance, logic and method, with a willingness to take intellectual risks."

Kober was indeed cautious and methodical, but she was also, as her hundreds of letters amply attest, funny, self-deprecating, charming, and intensely concerned about practically everything. She moved through her short life with a quiet, burning ardor—for teaching, for learning, for the just treatment of her fellow human beings—that belied her prim exterior and seemed born of what she evocatively called "a feeling for the fitness of things." Her correspondence also makes clear that she did allow herself to entertain, privately, some intellectually risky approaches to the riddle of Linear B. Some of these, arrived at independently by Ventris after Kober's death, would bring about its solution.

The scholarly field on which Kober did battle in the 1930s and '40s was very much a man's world, and it is understandable, if now unpalatable, that her male contemporaries so often characterized her in terms of maidenish qualities. That at least some twenty-first-century writers continue to accept this appraisal is far less understandable, and far less palatable.

In focusing primarily on Kober's story, I in no way intend to diminish the stunning achievement of Ventris, or of Arthur Evans, the English archaeologist who uncovered the tablets. It is simply that other writers have already recounted their work in some detail; these sources can be found in the references on page 289. Kober's role in the decipherment, so vital yet so long overlooked, is the logical narrative armature on which to build this book. I have chosen to quote extensively from her letters in the chapters devoted to her life, for it is through them, even more than through her masterful published writings, that she truly reveals herself.

All this said, *The Riddle of the Labyrinth* also discharges a debt to Ventris. In my daytime life, I have the great privilege of working as an obituary news writer at the *New York Times,* where I am paid to write the narrative histories of extraordinary people who have done extraordinary things. In September 1956, after Ventris died, obituaries appeared in newspapers throughout Europe. But for unknown reasons, most American papers, including the *Times,* overlooked the news of his death entirely. It can happen. Receiving timely news from abroad was a less reliable proposition then, and obituaries were less valuable journalistic properties. Assuming that word of the death reached the *Times*'s newsroom at all, it would have taken little more than one bleary-eyed night editor who had heard neither of Ventris nor of Linear B for the obituary to have been consigned to the spike. As a result, Ventris's achievement is far less well known to American readers than it might be. And so, to rectify the omission six decades belatedly—and to uphold the honor of my profession—here, too, is his story.

What is more, the process by which Ventris cracked the code has remained something of a black box all these years. As his biographer Andrew Robinson has astutely written: "There is no thread like Ariadne's running through the Linear B decipherment labyrinth. Even Ventris himself was unable to produce a coherent narrative of his method." By examining the architecture of the sturdy methodological bridge that Kober built, *The Riddle of the Labyrinth* is able to illuminate the steps Ventris took in his triumphant crossing.

If the course of the decipherment were charted on paper, the Kober and Ventris narratives would form two sides of an equilateral triangle, Kober's side slanting upward to the apex, and Ventris's, in mirror image, slanting downward from the apex to close the figure. But there is a third side—the base—and it represents the third actor in the drama, the charismatic Victorian archaeologist Arthur Evans, who unearthed the tablets in 1900.

More than any other investigators, it is these three, Evans, Kober, and Ventris—the digger, the detective, and the architect—who animate the decipherment, and it is to each of them in turn that this book's three major sections are devoted. And so it is with Evans, the foundation, that our story begins.

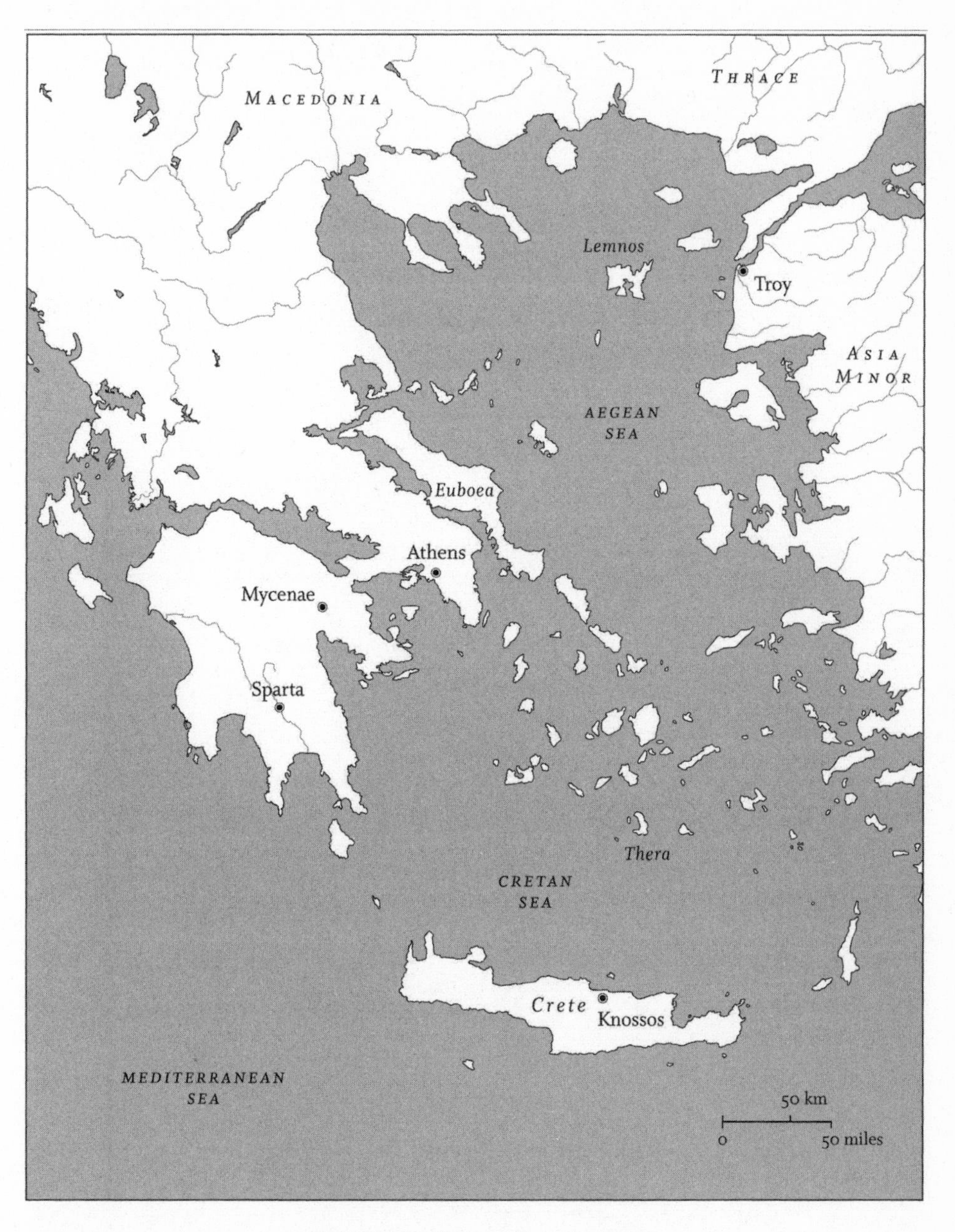

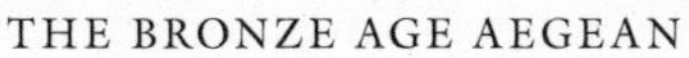
THE BRONZE AGE AEGEAN

PROLOGUE

BURIED TREASURE

Knossos, Crete, 1900

THE TABLET, WHEN IT EMERGED from the ground, was in nearly perfect condition. A long, narrow rectangle of earthen clay, it tapered toward the ends, resembling a palm leaf in shape. One end was broken: That was not surprising, after three thousand years. But the rest of the tablet was intact, and on it, inscribed numbers were plainly visible. Alongside the numbers was a series of bewildering symbols, which looked like none ever seen.

In the coming weeks, workmen would lift from the earth dozens more tablets, some fractured beyond repair, others completely undamaged. All were incised with the same curious symbols, including these:

The tablets were what Arthur Evans had come to Crete to find. It had taken him only a week to locate the first one, but his discovery would forever change the face of ancient history.

* * *

ON MARCH 23, 1900, Evans, a few carefully chosen assistants, and thirty local workmen had broken ground at Knossos, in the wild countryside of northern Crete near present-day Heraklion. There, not far from the sea, on a knoll bright with anemones and iris, Evans had vowed years earlier that he would dig.

He was rewarded almost immediately. Even before the first week was out, his workmen's spades turned up fragments of painted plaster frescoes in still-vivid hues, depicting scenes of people, plants, and animals. Digging deeper, they found pieces of enormous clay storage jars that reassembled would stand tall as a man. Still farther down, they encountered rows of huge gypsum blocks, the walls of a vast prehistoric building.

Evans had come upon the ruins of a sophisticated Bronze Age civilization, previously unknown, that had flowered on Crete from about 1850 to 1450 B.C. Predating the Greek Classical Age by a thousand years, it was the oldest European civilization ever discovered.

At forty-eight, Arthur Evans was already one of the foremost archaeologists in Britain. His discovery at Knossos, which the newspapers swiftly relayed around the globe, would make him among the most celebrated in the world. For the sprawling building beneath the knoll, he soon concluded, was none other than the palace of Minos, the legendary ruler of Crete, who crops up centuries later in Homer's epic poems, the *Iliad* and the *Odyssey.*

As Classical Greek myth told it afterward, King Minos had presided over a powerful maritime empire centered at Knossos. He held court in a huge palace resplendent with golden treasures and magnificent works of art, oversaw a thriving economy, and controlled the Aegean after making its waters safe from piracy. He was said to have installed an immense mechanical man, known as Talos and made of bronze, to patrol the Cretan shore and hurl rocks at approaching enemy ships.

It was for Minos, legend held, that the architect Daedalus had built the Cretan labyrinth, which housed at its center the fearsome Minotaur—half-man, half-bull. And it was Minos's daughter, Ariadne, with her ball of red thread, who helped her lover, Theseus, escape from the labyrinth, where he had been sent to be sacrificed. As Evans's prolonged excavation would reveal, the palace at Knossos spanned hundreds of rooms linked by a network of twisting passages. Surely, he would write, this vast complex was the historic basis of the enduring myth of the labyrinth.

Unseen for nearly three thousand years, the Knossos palace was hailed as one of the most spectacular archaeological finds of all time, "such a find," Evans wrote, "as one could not hope for in a lifetime or in many lifetimes." In his first season alone, he uncovered an exquisite marble fountain shaped like the head of a lioness, with eyes of enamel; carvings of ivory and crystal; ornate stone friezes; and, still more impressive, a carved alabaster throne, the oldest in Europe.

But these treasures paled beside what Evans found on the excavation's eighth day. On March 30, a workman's spade dislodged the first clay tablet. On April 5, a whole cache of tablets, many in perfect condition, was found in a single room of the palace.

The tablets, when Evans unearthed them, were Europe's earliest written records. Inscribed with a stylus when the clay was still wet, they dated to about 1450 B.C., nearly seven centuries before the advent of the Greek alphabet. The characters they contained—outline drawings in the shape of human figures, swords, chariots, and horses' heads, among other tiny pictograms—resembled the symbols of no known alphabet, ancient or modern.

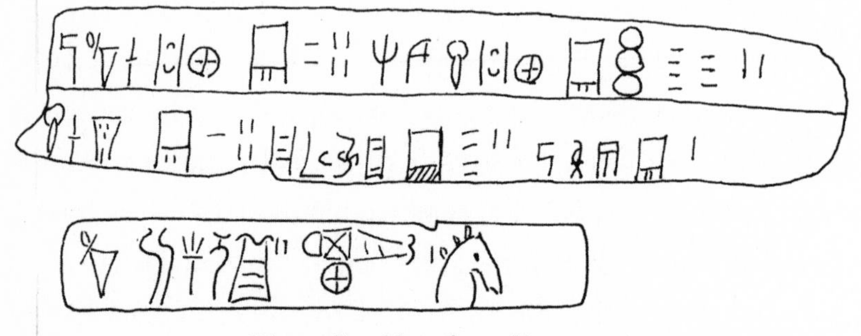

Linear B tablets from Knossos.

Evans named the ancient writing Linear Script Class B—Linear B, for short. (He also turned up evidence of a somewhat older Cretan script, likewise based on outline drawings, which he called Linear Script Class A.) By the end of his first season's dig, he had unearthed more than a thousand tablets written in Linear B.

Though Evans couldn't read the tablets, he immediately surmised what they were: administrative records, carefully set down by royal scribes, documenting the day-to-day workings of the Knossos palace and its holdings. If the tablets could be decoded, they would open a wide portal onto the daily life of a refined, wealthy, and literate society that had thrived in Greek lands a full millennium before the glory of Classical Athens. Once their written records could be read, the Knossos palace and its people, languishing for thirty centuries in the dusk of prehistory, would suddenly be illuminated—with a single stroke, an entire civilization would *become* history.

But which civilization was it? As Evans well knew, many ethnic groups had passed through the Bronze Age Aegean, and there was no way to tell whose language, and whose culture, Linear B represented. To him, though, this seemed a small impediment. Evans was already something of an authority on ancient scripts, and with characteristic assurance, he assumed he would one day decipher this one. By 1901, only a year after the first tablet was unearthed, he had commissioned Oxford University Press to cast a special font, in two different sizes, with which to typeset the Cretan characters.

But Evans underestimated the formidable challenge Linear B would pose. An unknown script used to write an unknown language is a locked-room mystery: Somehow, the decipherer must finesse his way into a tightly closed system that offers few external clues. If he is very lucky, he will have the help of a bilingual inscription like the Rosetta stone, which furnished the key to deciphering the hieroglyphs of ancient Egypt. Without such an inscription, his task is all but impossible.

As Evans could scarcely have imagined in 1900, Linear B would become one of the most tantalizing riddles of the first half of the twentieth century, a secret code that defied solution for more than fifty years. As the journalist David Kahn has written in *The Codebreakers,* his monumental study of secret writing, "Of all the decipherments of history, the most elegant, the most coolly rational, the most satisfying, and withal the most surprising" was that of Linear B.

The quest to decipher the tablets—or even to identify the

language in which they were written—would become the consuming passion of investigators around the globe. Working largely independently in Britain, the United States, and on the European continent, each spent years trying to tease the ancient script apart. The best of them brought to the problem the same meticulous forensic approach that helps cryptanalysts crack the thorniest codes and ciphers.

No prize was offered for deciphering Linear B, nor were the investigators seeking one. For some, like Evans, the chance to read words set down by European men three thousand years distant was compensation enough. For others, the sweet, defiant pleasure of solving a cryptogram many experts deemed unsolvable would be its own best reward.

Today, in an era of popular nonfiction that professes to find secret messages lurking in the Hebrew Bible, and of novels whose valiant heroes follow clues encoded in great works of European art, it is bracing to recall the story of Linear B—a real-life quest to solve a prehistoric mystery, starring flesh-and-blood detectives with nothing more than wit, passion, and determination at their disposal.

Over time, two besides Evans emerged as best equipped to crack the code. One, Michael Ventris, was a young English architect with a mournful past, whose fascination with ancient scripts had begun as a boyhood hobby. The other, Alice Kober, was a fiery American classicist—the lone woman among the serious investigators—whose immense contribution to the decipherment has been all but lost to history. What all three shared was a ferocious intelligence, a nearly photographic memory for the strange Cretan symbols, and a single-mindedness of purpose that could barely be distinguished from obsession. Of the three, the two most gifted would die young, one under swift, strange circumstances that may have been a consequence of the decipherment itself.

Considered one of the most prodigious intellectual feats of modern times, the unraveling of Linear B has been likened to Crick and Watson's mapping of the structure of DNA for the magnitude of its achievement. The decipherment was done entirely by hand, without the aid of computers or a single bilingual inscription. It was accomplished, crumb by crumb, in the only way possible: by finding, interpreting, and meticulously following a series of tiny clues hidden within the script itself. And in the end, the answer to the riddle defied everyone's expectations, including the decipherer's own.

To Ventris, the solution brought worldwide acclaim. But before long it also brought doubt, despair, personal and professional ruin, and, some observers believe, untimely death.

All this was decades in the future that March day at Knossos, when the first brittle tablets emerged from the ground. But of one thing Arthur Evans was already certain. Guided by the smallest of clues, he had come to Crete in search of writing from a time before Europe was thought to *have* writing. And there, he now knew beyond doubt, he had found it.

BOOK ONE

The Digger

Arthur Evans at Knossos, 1901.

1

THE RECORD-KEEPERS

EVANS CAME TO CRETE TO fill a void. In 1876, Heinrich Schliemann, a wealthy German businessman with a burning interest in the classics, began excavating a site on the Greek mainland, about seventy miles southwest of Athens. The site he chose was fabled as the home of Mycenae, the ancient city known from Homer as the kingdom of Agamemnon, brother-in-law of the beautiful Helen of Troy.

As Evans would on Crete a quarter century later, Schliemann unearthed the relics of an advanced Bronze Age civilization, this one lying two hundred miles north of Crete, over the sea. Mycenae had been a real, prosperous, well-run society that flourished in the second millennium B.C. Before long, the most visible Bronze Age ruins there could be dated to about 1600 to 1200 B.C.: In the late 1880s and afterward, the distinguished archaeologist Flinders Petrie uncovered Mycenaean trade goods, including ceramic vessels, while excavating Egyptian sites of known date.

Schliemann was already famous for unearthing vanished worlds. In the early 1870s, he had uncovered what he believed to be Troy itself, at Hisarlik, in present-day Turkey. There, where King Priam was fabled to have reigned, and where a long, bloody war was said to have raged after Priam's son, Paris, abducted Helen from her home in Sparta, Schliemann dug fruitlessly for several years. Shortly before the dig was to end, he later wrote, he came upon a golden hoard: gold diadems and goblets and buttons and earrings and rings. It would be known ever after as the treasure of Priam.

Schliemann's excavation methods, which involved the wholesale hacking away of huge, potentially fruitful layers of soil, distress many modern archaeologists. Over the years, the authenticity of some of his finds, both at Troy and Mycenae, has been questioned. Today, some critics view him more as tomb robber than archaeologist.

What Schliemann's archaeology lacked in scientific rigor it amply made up in romantic fervor. Driving him to dig at both sites was the desire to prove that Homer's epic poems, the *Iliad* and the *Odyssey,* were factual works of history. (The poems are now thought to have been composed in the eighth or seventh century B.C. Schliemann's earnest belief that they were nonfiction is one that few scholars, now as then, have been inclined to share.)

But Schliemann's work remains important for having taken civilizations thought to be the figments of tale-tellers and placed them, at least possibly, within the realm of history. At Hisarlik, he helped animate the heroes of the Trojan War, fought, some sources say, in the thirteenth or twelfth century B.C. At Mycenae, too, he brought the world of the Aegean Bronze Age to light, showing that a high civilization was already in full flower on the Greek mainland a thousand years before Classical times.

Ever since Schliemann dug at Mycenae, the span of early Greek history from the sixteenth to the thirteenth centuries B.C.—the era when this mainland kingdom was at its height—has been known as the Mycenaean Age.

Mycenae had been a walled citadel. It was made of stone blocks so massive, the scholar John Chadwick wrote, that "the later Greeks understandably concluded that the walls had been built by giants." Still standing in Schliemann's day was the city's famed Lion Gate, a portal of enormous blocks topped with two lions in carved relief. It was a splendid feat of engineering, yet so different stylistically from the triangular pediments and fluted columns of Classical Greece. Digging down into the circle of deep "shaft graves" inside the city's walls, Schliemann brought up priceless golden artifacts, engraved gems, silver vessels, and other treasures.

But just as striking was what Schliemann *didn't* find. Despite the great refinement of the Mycenaean kingdom, despite the well-oiled administration that had clearly sustained it, there was no sign of writing anywhere. Though Schliemann excavated on a typically grand scale, pouring his vast personal resources into the dig, he turned up no clay tablets, no inscriptions cut in stone—no evidence whatsoever that this sophisticated mainland society had been literate.

This bothered Evans. Like many people, Arthur Evans had followed the newspaper accounts of Schliemann's dig with rapt interest. Surely, Evans thought, so advanced a civilization, with all its attendant bureaucracy, would have had a means of keeping written records. Mycenaean society was far too complex—too *competent*—he believed, not to have been acquainted with writing in some form. As Evans would later explain, his Victorian worldview on unabashed display, "It seemed incredible that [such] a civilisation . . . could in the department of writing, have been below the stage attained by Red Indians."

Perhaps the Mycenaeans had written on perishable materials, like palm leaves, bark, or hide. But by the Victorian era, beguiling hints that they had written on sturdier stuff had begun to surface. In the early 1890s, Chrēstos Tsountas, an associate of Schliemann working at Mycenae, unearthed a clay amphora—a type of two-handled vessel—with three "linear" signs incised on one handle. Nearby, in a Mycenaean tomb, Tsountas unearthed a stone vase whose handle was engraved with four or five linear symbols. Elsewhere on the mainland, linear signs were found painted on a few pieces of pottery.

The three linear signs incised on the handle of a mainland vessel. The leftmost symbol was also seen at Knossos.

There were also hints on Crete. In the late 1870s, the remains of a Bronze Age wall were exposed at Knossos, on the site at which Evans eventually dug. In the early 1880s, one of the wall's huge gypsum blocks was discovered to be inscribed with a series of linear symbols, which scholars quickly dismissed as "masons' marks." Strikingly, the wall at Knossos and the amphora at Mycenae, separated by two hundred miles of sea, had a symbol in common, the character . The more Evans considered the question, the more he became convinced that some form of writing had existed in the Aegean in Mycenaean days.

Evans was only in his twenties when Schliemann dug at Mycenae, but he already possessed the characteristics necessary for a world-class archaeologist: tirelessness, fearlessness, boundless curiosity, wealth, and myopia. By the 1890s, when he began to attack the problem in earnest, he had already traveled to remote corners of the globe; become a recognized expert on ancient coins; lived for long periods in the Balkans, where he was a passionate advocate of the Slavic nationalist cause; and been appointed keeper, as the head curator was known, of the Ashmolean, the distinguished archaeological museum in Oxford. Armed with these unassailable assets, Evans went forth to find evidence of writing in Mycenaean times. Clue after clue would point him toward Crete.

Excavation was in Evans's blood. His father, Sir John Evans, a wealthy paper manufacturer, was also a distinguished amateur geologist, archaeologist, and numismatist. Known among his contemporaries as Evans the Great, John Evans "helped to lay the foundations of modern geology, paleontology, anthropology, and archaeology despite the fact that he could dedicate only Sundays and holidays to the dim past," Sylvia L. Horwitz wrote in *The Find of a Lifetime,* her biography of Arthur Evans.

The eldest of five children of John and Harriet Dickinson Evans, Arthur John Evans was born on July 8, 1851. He was reared with his siblings in the Hertfordshire countryside in a grand house overflowing with fossils, prehistoric stone tools, arrowheads, Roman coins, and ancient pottery, the stuff of their father's weekend trade. Arthur was a sober, curious boy who could spend hours intently studying old coins, though he seemed indifferent to conventional book-learning. (Because Arthur had not mastered Latin grammar by the age of six, as his father before him had done, his paternal grandmother confided to Harriet her fear that the child was "a bit of a dunce.")

On New Year's Day 1858, when Arthur was six and a half, Harriet Evans died after giving birth to her fifth child. Afterward, as Arthur's much younger half sister, Dame Joan Evans, recounted in her biography of him, *Time and Chance,* "John Evans wrote in his wife's diary that [the children] did not seem to feel her loss; more than seventy years later Arthur Evans was to write an indignant *NO* in the margin." The next year, John Evans married a cousin, Fanny Phelps, who was by all accounts a loving stepmother to Harriet's children.

As a schoolboy at Harrow, Arthur won prizes in natural history, modern languages, and the writing of Greek epigrams. Entering Oxford, he studied history, graduating with first-class honors in 1874. As a twenty-year-old undergraduate, he published his first scholarly article, "On a Hoard of Coins Found at Oxford, with Some Remarks on the Coinage of the First Three Edwards." It was his first public foray into the antiquarian circles of which his father was a leading light. (As a result of his early work, Arthur became known in those circles as "Little Evans, son of John Evans the Great," a description that must surely have rankled.) After graduating from Oxford, Arthur studied briefly in Germany before striking out on the first of several long trips to the Balkans. The region interested him intensely, and he would live and work there for much of the next decade.

At the time, the Balkans were under the control of the Ottoman Empire, and the Slavic peoples of the region were eager to throw off its yoke. Evans became a staunch public champion of the Slavs' struggle for self-determination—a struggle that was often violent during the years he was there. In impassioned, unapologetically partisan prose, he filed a series of dispatches to the *Manchester Guardian* chronicling the heroism of the Slav resistance fighters. From his eventual base in Ragusa—the old Italian name for Dubrovnik, in what is now Croatia—Evans ranged over the remote corners of Serbia, Bosnia, and Herzegovina by foot, horseback, and steamer. He traversed the wild countryside to investigate reports of Turkish atrocities in remote villages, stripped off his clothes to ford icy rivers, and scaled cliffs to meet with fierce Turkish overlords in their mountaintop command posts. He was often uncomfortable, usually inconvenienced, and occasionally imprisoned. None of this seemed to bother him very much.

In 1876, when he was twenty-five, Evans published the first of his two books on the Balkans, *Through Bosnia and the Herzegóvina on Foot,* whose grandiloquent full title, *Through Bosnia and the Herzegóvina on Foot during the Insurrection, August and September 1875: With an Historical Review of Bosnia, and a Glimpse at the Croats, Slavonians, and the Ancient Republic of Ragusa,* left little doubt as to the sweep of his enterprise. The exploits he recounted gave one normally unflappable reader cause for concern. "Mind where you travel!" the intrepid British explorer Richard Burton wrote Evans in 1877, after reading his work.

In September 1878, Evans married Margaret Freeman. Small, plain, intelligent, and three years older than he, she was a daughter of Edward Augustus Freeman, a well-known historian now best remembered for his fiercely held views on Aryan racial supremacy. After their marriage, Evans brought Margaret to live in his beloved Ragusa, where they occupied a pleasant house by the sea in the old walled city. Evans had horrified his father by signing a twenty-year lease on the place.

He would not get to stay nearly so long. In 1882, with the region now controlled by the Austro-Hungarian Empire, Evans was arrested for his political activities. Released by Austrian officials after seven weeks in a local jail, he was expelled from Ragusa. With Margaret, he returned to England. In 1884, Evans was appointed keeper of the Ashmolean, and the couple settled in Oxford.

By this time, Arthur Evans embodied the Victorian age writ large, or, more precisely, writ small. A diminutive man of barely five feet, Evans possessed all of his era's thirst for scientific inquiry, most of its grand passions, and many of its reflexive prejudices. Deeply curious about far-off lands and their people, he nonetheless bristled when Bosnian villagers addressed him as *brat,* "brother." An ardent defender of the downtrodden, he could also write, as he did in *Through Bosnia and the Herzegóvina on Foot*: "I don't choose to be told by every barbarian I meet that he is a man and a brother. I believe in the existence of inferior races, and would like to see them exterminated." (Evans tempered his verdict slightly in the next sentence, writing, "But . . . it is easy to see how valuable such a spirit of democracy may be amongst a people whose self-respect has been degraded by centuries of oppression.")

Despite his small stature, Evans cut an imposing figure, always impeccably turned out in suit, tie, vest, and hat, and carrying a sturdy walking stick he had nicknamed Prodger. Prodger was not so much to lean on as to see by: From earliest childhood, Evans had been desperately nearsighted. As Joan Evans wrote: "His short sight, for which he refused to wear proper glasses, made him carry his head in a rather peering way. Moreover, he suffered from an extreme degree of night blindness, so that in the winter terms at Harrow he needed a friendly guide to steer him to or from afternoon school."

But Evans's myopia, so constraining in other ways, proved a staggering advantage in his line of work. Unlike most people, he could see things with near-microscopic precision at extremely close range. When he was a boy, his stepmother, Fanny, wrote fondly of his peering at the face of an old coin "like a jackdaw down a marrow bone."

As Evans squinted at a coin, engraved gem, or other tiny artifact held directly before his eyes, he could make out fine details many other experts missed. It is fair to say that had Arthur Evans not been so terribly myopic, the Linear B tablets at Knossos would not have been found when they were. For a series of clues so minute that only he could interpret them had told him where to dig.

BY THE MID-1880s, the Greek Bronze Age had begun to exert its hold on Evans. After his expulsion from the Balkans, he chafed in England and was soon beset by his constitutional wanderlust. He could not go back to Ragusa. Instead, in 1883, he journeyed with Margaret to Greece.

In Athens, the Evanses called on Schliemann himself. Now in his early sixties, Schliemann lived there in profuse splendor with his beautiful young Greek wife, surrounded by his glittering spoils. He regaled the couple with tales of his digs and showed them some of his finds, including gold jewelry and small, beadlike gemstones engraved with naturalistic designs.

As a result of the visit, and the five months he and Margaret spent traveling in Greece afterward, Evans grew fascinated by the Mycenaean world.

In the Victorian age, the widely accepted view of Greek history was that it had begun in 776 B.C., the date of the first known Olympics. The Greek alphabet, borrowed from the Phoenicians not long before, had made writing possible, and with it, recorded history. The Classical Era, with its spectacular achievements in arts, letters, and science, would follow soon afterward, at its height spanning the seventh to the fourth centuries B.C. Before the Classical Era, historians believed, lay a long Greek Dark Age. Lasting from about 1200 to 800 B.C., it was a time in which literacy, high art, and skilled architecture were unknown in Greek lands. By Homer's day, circa 800 B.C., Greece was "at a comparatively low level of civilization," as John Chadwick wrote in 1976. And yet, he added, the Greece Homer *describes* in his epics—the Greece of five centuries earlier—"is a network of well-organized kingdoms capable of joint military action; its kings live in luxurious stone-built palaces, adorned with gold, ivory and other precious metals."

Homeric epics were composed and transmitted orally: There was, after all, no alphabet with which to write them down. Yet in those epics, as Chadwick noted, Homer sang *about* writing:

> When Homer describes a letter entrusted to a traveller—it was, ironically, a request for the bearer to be quietly liquidated—Homer describes it as something exotic and almost magical; writing was no more than a dim memory. But some idea of the Mycenaean world could well have been passed down through the Dark Ages to Homer, and the

tradition of verse-making may go back to the Mycenaean palaces.

Nineteenth-century scholars dismissed Homer's accounts of Bronze Age life as pure poetic fancy. The glories of Classical Greece, the strong implication went, had sprung full blown from the long cultural vacuum that preceded them.

Unlike most historians, Arthur Evans had grown up with his hands in the grit of prehistory. When he was eight, his father and two colleagues had unearthed Stone Age implements from a gravel pit in the Somme River valley in France. In so doing, as Joseph Alexander MacGillivray wrote in *Minotaur*, his life of Arthur Evans, they helped prove to the religious and scholarly communities "that human beings had lived on this earth for a far greater time than the clerics had allowed for." As an older boy, Arthur often accompanied his father on digs; while studying in Germany, he undertook a dig of his own at a Roman site in Trier.

To Evans, the idea that Classical Greece had arisen out of nowhere was absurd. It was plain to him that Greek civilization, like any other, had *come from somewhere*—an idea Schliemann's Mycenaean discoveries only served to reinforce. Returning from Greece to Oxford, he began to think deeply about the Mycenaeans and what their influence on the Greek Classical Age might have been.

Schliemann's dig at Mycenae had peeled back layers of time, exposing a community that had thrived during the Aegaen Bronze Age. Mycenae was clearly a high civilization, with beautiful art and impressive architecture. Yet it seemed to have no writing. "Such a conclusion," Evans flatly declared, "I could not bring myself to accept."

Who were the Mycenaeans and where had they come from? What language did they speak? The gold and jewels Schliemann unearthed could hint at the Mycenaean way of life, but in the end they were mute testaments. Without written records, Evans knew, it would be impossible to learn much more. "The discoveries of Schliemann revealed so high a type of civilisation in the prehistoric Aegean, that if writing had proved to be unknown it would have been its absence which would have called for explanation," he later wrote. Evans resolved to go in search of it, though he would not be able to turn his full attention to the quest till the end of the century.

In Oxford, meanwhile, Evans was occupied with transforming the Ashmolean from a haphazard cabinet of curiosities into a world-class museum of art, archaeology, and antiquities. The keepership allowed for frequent travel, and he spent much time abroad, scouring Europe for artifacts to add to the collection. He was also busy orchestrating a suitable home for himself and Margaret. Evans had bought sixty acres on a hill outside Oxford, with commanding views of the countryside; it was a place he had loved in his student days. There he would build his house, which he named Youlbury, "from the ancient name of the heath below," as Horwitz wrote.

For Evans, there was a sense of urgency about the project: In 1890, Margaret had been diagnosed with tuberculosis, and he hoped Youlbury's clean Oxfordshire air would restore her to health. But she did not live to see it finished. In March 1893, Margaret Evans died after fifteen years of marriage, leaving Arthur a widower at forty-one. "For the rest of his life he wrote on black-edged paper," Horwitz wrote. "Even his scribbled notes were bordered in black." Youlbury, a sprawling Victorian behemoth, was completed the next year, and Evans moved into it alone.

Even before Margaret's death, Evans had begun to explore the idea that writing was used in the Mycenaean world. In February 1893, just weeks before she died, he returned to Athens, where he picked over small, dusty treasures in antiquities shops. What he found there would eventually lead him to Knossos: small, prism-shaped stones of three and four sides, often of semiprecious material like red or green jasper, carnelian, or amethyst, pierced for wearing. Each face of the stone was engraved, Evans wrote, with "a series of remarkable symbols." These symbols—ornate hieroglyphic pictures of people, animals, and objects—were, as he observed, "not a mere copy of Egyptian forms."

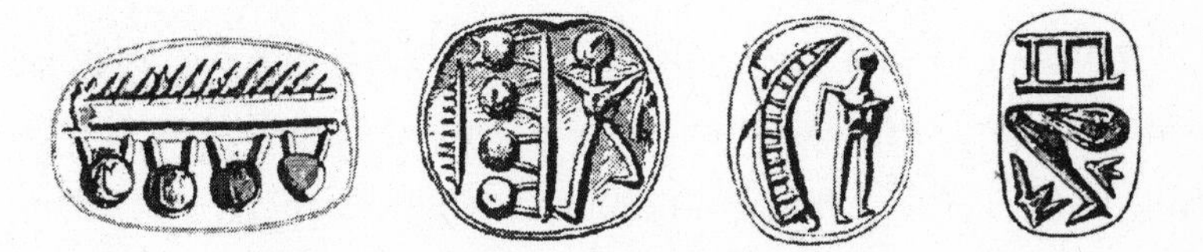

Cretan stones with hieroglyphic engravings, acquired by Arthur Evans.

The stones he had come upon are called seal-stones. Designed to make an impression in soft clay or wax, they were a means of marking ownership in prehistoric times. They reminded Evans of something Schliemann had shown him: the small, beadlike gems unearthed at Mycenae, also pierced and engraved with tiny stylized symbols. (Schliemann's bead gems, however, were strictly ornamental.)

In Athens, Evans bought as many seal-stones and engraved gems as he could find. With each purchase, he asked the dealer where the stones were from. The answer was nearly always the same: They had come from Crete. "To Crete," Evans wrote, "I accordingly turned."

TO MODERN OBSERVERS, Crete seems merely a seaborne extension of Greece; in fact, it did not become part of the Greek state until 1913. The largest of the Greek islands, it lies almost equidistant between Europe, Asia, and North Africa, for centuries a handy stopping point for Mediterranean seafarers. As archaeologists of Evans's time were already aware, Crete's earliest known inhabitants were unrelated to the Greeks who would later people the mainland. "It was clearly recognized by the Greeks themselves," Evans wrote, "that the original inhabitants of Crete were 'barbarian' or un-Greek."

In the centuries to come, Crete was repeatedly settled, invaded, colonized, traded with, resettled, reinvaded, and recolonized. By 1900, when Evans began digging there, the island was a web of ethnic, linguistic, and cultural influences stretching back thousands of years.

Evans paid his first visit to Crete in March 1894. The island was then part of the Ottoman Empire, and within days of his arrival, he was plunged unwittingly into the hostilities between Greek Christians and Turkish Muslims there. As he wrote in his journal on March 17:

> In the evening some excitement. Knowing the straight road, I walked back at 9.45 in clear moonlight from the chief café to the inn. Hardly in my room, than three Christians burst in to the inn to say that two Turks had followed

> me to assassinate me, and would have stabbed me if they had not come after them. . . . People seem excited about it, but what is certain is that I was not.

In the following days, as he roamed the island's rocky country by foot and mule, Evans came upon seal-stones and engraved gems like those he had bought in Athens, carved with the same curious symbols. Many were owned by Cretan peasant women, who prized them as amulets. Known locally as *galopetras* ("milk stones"), they were worn by nursing mothers, who believed they ensured a plentiful supply of breast milk. Evans bought as many of the stones as he could; if a young mother refused to sell her charm, he could often persuade her to let him make a rubbing of it.

The engravings on the stones, Evans quickly came to believe, were no mere decorations. They were too stylized for that, and too systematic. The carved symbols often occurred in clearly defined groups, and the same symbols might recur again and again on different stones. The carvings clearly *signified* something very particular to the Bronze Age people who had made them. "It is impossible to believe that the signs on these stones were simply idle figures carved at random," he wrote in 1894. "Had there not been an object in grouping several signs together it would have been far simpler for the designer to have chosen single figures or continuous ornament to fill the space at his disposal."

Evans knew that he had come upon a system of written communication, used long before the Phoenicians invented the alphabet in the eleventh century B.C. and longer still before the alphabet made its way to Greece at the end of the ninth.

It was a written record of the sort Schliemann had expected to find at Mycenae. And now Evans had found it elsewhere in the Aegean, dating to Mycenaean times. The Cretan stones, he later wrote, offered clear evidence "that the great days of the island lay beyond history."

By the end of 1893, before he had even set foot on Crete, Evans had felt sure enough of the markings on his Athenian seal-stones to announce his discovery in public, declaring to a London audience that he possessed "a clue to the existence of a system of picture-writing in the Greek lands." In 1894, after he returned from the island, he published his first significant article about the engraved Cretan stones. In it, he argued that "an elaborate system of writing did exist within the limits of the Mycenaean world."

Evans identified two types of Cretan writing. On some stones, the carvings were clearly hieroglyphic, teeming with pictograms of people, plants, and animals. On others, the symbols were "linear and quasi-alphabetic," as if the hieroglyphs had been reduced to their clean, bare outlines in "a kind of linear shorthand." "Of this linear system too," he wrote prophetically in 1894, "we have as yet probably little more than a fragment before us." What was needed, Evans knew, was a full-scale excavation on Cretan soil. Over the next few years, he paid repeated visits to the island, eventually choosing Knossos as the place to dig.

It was no random selection. Tradition held that Knossos had been the chief city of Cretan antiquity, the fabled seat of Minos's empire. "Broad Knossos," Homer had called it in the *Iliad*. In the *Odyssey,* he sang:

One of the great islands of the world
in midsea, in the winedark sea, is Krete:
spacious and rich and populous, with ninety
cities and a mingling of tongues. . . .
And one among their ninety towns is Knossos.
Here lived King Minos whom great Zeus received
every ninth year in private council.

By Evans's day, archaeologists had already unearthed significant finds on the island. At the time, the part of Crete where Knossos was thought to have stood was known locally as Kephala. (The name was a shortened form of the half-Greek, half-Turkish phrase *tou Tseleve he Kephala,* "squire's Knoll.") In 1878, a Greek linguist with the historically evocative name of Minos Kalokairinos brought twenty workmen to Kephala and started digging. He found the remains of a vast building made of gypsum blocks, whose rooms were filled with huge ceramic jars.

Three weeks into the dig, the Cretan Assembly ordered Kalokairinos to stop work. As MacGillivray wrote, "The reasoning, which Kalokairinos accepted, was that he might begin to reveal the sort of enviable artifacts that would almost certainly be removed from Crete to Constantinople." Word of the discovery did get around in archaeological circles, and modern historians often credit Kalokairinos as the first true discoverer of the Palace of Minos. In the early 1880s, William James Stillman, a former American consul on Crete, examined a wall that Kalokairinos had exposed before the dig was halted and noticed the curious "masons' marks" carved into the stone.

Schliemann, too, had his eye on Kephala. Starting in 1883, he determined to excavate the Palace of Minos himself: It would be the final triumph, he hoped, of his storied career. Kephala was owned by an extended Turkish family, and though Schliemann tried to secure permission to dig there, he was unable to do so before he died in 1890. Schliemann's death left the way open for Evans. "Nor can I pretend to be sorry that he did not dig at Knossos," Evans would write years later.

Besides the seal-stones and engraved gems he had already secured on Crete, Evans had encountered something even more exciting. In 1895, a local man showed him something he had found on the ground at Kephala: a "slip" of burned clay, about the size and shape of a slip of paper, incised with linear signs that "seemed to belong to an advanced system of writing," as Evans said. He added, decisively: "On the hill of Kephala . . . I resolved to dig"

Needing digging rights, Evans did what any self-assured Victorian of means would do: He simply bought the property. But the process turned out not to be so simple, even for a man of his wealth and determination. Schliemann had made two fortunes, first by starting a bank in Sacramento amid the California Gold Rush and later by cornering the European indigo market, yet even he had had no luck on Crete.

In 1894, after much negotiation with Kephala's owners, "native Mahometans, to whose almost inexhaustible powers of obstruction I can pay the highest tribute," Evans managed to buy a quarter-share of the property for 235 British pounds. This gave him the right to force the sale of the remaining three-quarters. But over the next few years, a bloody insurrection on the island, in which the Greek Cretans tried to rout their Turkish oppressors, made further negotiation impossible. As he had done in the Balkans, Evans threw his support behind the local people in their fight to break free of the Ottoman Empire.

Though he could not yet begin to dig, Evans was certain of the deep importance of what he had already found on Crete. The seal-stones and engraved gems he had obtained from the island's peasant women were, he wrote in 1897, "striking corroboration" that "long before our first records of the Phoenician alphabet, the art of writing was known to the Cretans."

The insurrection raged on for several years before the Greeks prevailed; the last of the Turkish forces left the island in late 1899. The next year, "after encountering every kind of obstacle and intrigue," Evans bought the rest of Kephala for 675 pounds. After extensive preparations in England—he equipped himself with a gross of nail brushes, two dozen tins of ox tongue, twenty tins of sardines, twelve plum puddings, a case of Eno's Fruit Salt (a stomach remedy), a Union Jack, and a fleet of iron wheelbarrows, among other things—he landed on Crete in early March 1900. There he set about disinfecting and whitewashing the rented house in which he and his assistants would live.

On March 23, with the Union Jack flying imperially overhead, Evans broke ground at Kephala. Excavating the site and restoring it to its former glory would occupy him till the end of his life.

KEPHALA HAD BEEN inhabited as far back as the Stone Age. Digging deeply, Evans found stone implements and crude artifacts that dated to Neolithic times, about 6100 B.C. But it was the

Bronze Age he was after, in particular the span from about 1850 to 1450 B.C., when the Palace of Minos had flourished.

Though the palace's outer walls were made of great stone blocks, its infrastructure was wood, an engineering safety measure in an earthquake-prone region. As layer upon layer of charred timber revealed, the palace had been ravaged, burned, and rebuilt several times over the centuries. The cause of these repeated destructions could only be guessed at: Any one of them could have resulted from an earthquake, a lightning strike, or a sacking at the hands of an invading enemy. What was clear was that at the start of the fourteenth century B.C., in some final catastrophe, the palace was sacked and burned one last time. It was rebuilt and partly reoccupied, but Knossos would never again be a seat of power.

Under Evans's direction, years of excavation would reveal a building larger than Buckingham Palace, spread over six acres. Like the ruins of Mycenae, the Palace of Minos could be dated with reference to Egyptian trade goods: In one of the site's deeper layers, Evans's workmen turned up a small Egyptian statue, carved of diorite and known to date from about 2000 B.C.

The palace comprised hundreds of rooms and what had been multiple stories, grand staircases, vast halls, storerooms, and artisans' workshops. Evans's men uncovered the remains of a sophisticated hydraulic system that had included terra-cotta pipes, bubbling fountains, bathtubs, and even toilets that could be flushed with water. By the end of the 1900 season, which lasted nine weeks, the initial group of 30 workmen had grown to about 180. (In the interest of promoting harmony on the island in the wake of the insurrection, Evans employed both Christian and Muslim workers.) The men dug and lifted and hauled; nearby, the women sifted the loose soil for tiny treasures, like beads and plaster fragments, which they carefully washed.

In the course of the season, Evans's workers unearthed an exquisite alabaster vase shaped like a triton shell; the high alabaster throne; pieces of many different frescoes, which Evans would arrange to have painstakingly restored; statuary; painted pottery; and the charred remains of graceful wooden columns that, like those at Mycenae, were smooth, round, and downward-tapering, wider at the top than at the bottom. So delighted was Sir John with his son's discoveries that he sent him 500 British pounds.

But all this, as Evans's assistant John Linton Myres would write, was "almost thrown into the shade" by the discovery that season of the inscribed clay tablets, "a discovery which carries back the existence of written documents in the Hellenic lands some seven centuries beyond the first known monuments of the historic Greek writing."

On March 30, workmen lifted from the soil of Knossos "a kind of baked clay bar, rather like a stone chisel in shape, though broken at one end, with script on it and what appear to be numerals," as Evans later wrote. He saw immediately that the linear script on the bar resembled that of the clay "slip" he had been shown in 1895. The slip had been destroyed not long afterward, a casualty of the Cretan insurrection, but Evans had had the foresight to make a copy of it.

The discovery of writing at Knossos was, Evans later wrote,

"the dramatic fulfillment of my most sanguine expectations." In the weeks that followed, his workmen unearthed more and more tablets, many badly broken but some completely intact. At times, they came upon entire small chambers filled with tablets, the archival record rooms of the Palace of Minos. On April 5, 1900, they encountered a terra-cotta bathtub full of tablets—they had fallen into the tub centuries before, when an upper story gave way. Also in the tub were bits of charred wood, suggesting the tablets had originally been stored in wooden boxes. This was confirmed elsewhere in the palace, when tablets were found next to a set of small bronze hinges—hinges that had held the wooden box lids in place. On May 10, Evans wrote his father that he had "just struck the largest deposit yet, some hundreds of pieces."

Like the clay bar Evans first unearthed at the site, most of the Knossos tablets were small and wedge-shaped, between two and seven inches long and a half inch to three inches high. Tapered at the ends, they were clearly designed to fit comfortably in a scribe's hand. They were made of ordinary local clay, incised with a stylus of some kind.

In addition, there were some larger, squarer tablets, also made of clay and sometimes fashioned around armatures of straw. Where the smaller tablets had room for just a line or two of writing, perhaps ten or twenty characters in all, the larger tablets held much more. One very large rectangular tablet, more than ten inches high and six inches wide, was incised with twenty-four lines of text. It would come to be known as the "Man" tablet for the column of "man" pictograms (𐂀) down its right-hand edge.

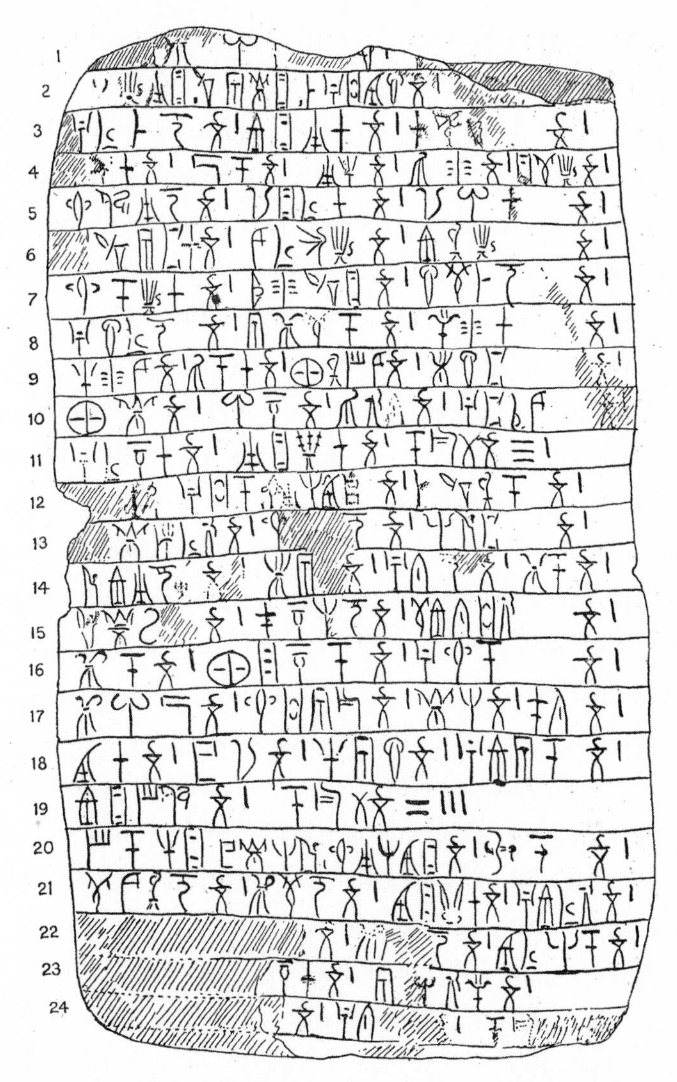

The twenty-four-line "Man" tablet from Knossos. The "man" symbol (𐂀) followed by a numeral, is repeated down its right-hand edge.

The scribes of Knossos were superb bureaucrats. Tablets were arranged by subject in their boxes, the file cabinets of the Bronze Age Aegean. Once filled and closed, each box was secured with a clay sealing indicating its contents. The sealing was impressed with a seal-stone like the ones Evans had encountered on his first trip to Crete. (Long before the Knossos tablets could be read, their subject matter—grain, livestock, chariots, weapons, and the like—could often be gleaned from the pictograms on the seals.)

Evans would identify three different scripts at Knossos. The first was a hieroglyphic script, the same kind he had seen on the seal-stones and engraved gems. From the depth of the soil in which traces of the script were found, he determined that it was in use from about 2000 to 1650 B.C. But while this script was found often on the clay sealings used to secure boxes of tablets, it was rare on the tablets themselves: In the whole of the palace, Evans came across only a single cache of tablets bearing hieroglyphic inscriptions.

Other tablets featured what he described as "a new system of linear writing," which had evolved from the hieroglyphic script in the eighteenth century B.C. This linear script was a "style of writing fundamentally different from that of the hieroglyphic class, and far ahead of it in development. . . . The letters themselves . . . are of a free, upright European character."

By 1902, Evans had further distinguished two types of linear script. The first, Linear Script Class A, was used from about 1750 to 1450 B.C. The second, Linear Script Class B, developed out of Linear A toward the end of this period. It was in use until the final destruction of the palace in the early fourteenth century B.C. Evans called the scripts "linear" not because their characters were arrayed in lines, although they were, but because those characters were made by means of linear strokes—a method quite different from the cuneiform writing of ancient Mesopotamia, in which signs were impressed in clay with a wedge-shaped tool. (With its unadorned outline forms, the Phoenician alphabet, which gave rise to nearly all the alphabets of the modern world, was also a linear script. So are its descendants, including our own familiar Roman alphabet.)

In contrast to Linear B tablets, Linear A tablets, like these, are nearly always unruled.

The great majority of the Knossos tablets were written in Linear B, including the first ones unearthed in the spring of 1900 and the large ones containing many lines of text. Linear B looked similar to Linear A but was by no means identical. The two scripts had many characters in common—among them 𐀳, 𐀚, 𐀅, 𐀊, 𐀯, 𐀏, 𐀇, 𐀭, 𐀦, and 𐀣—but each also used characters not found in the other. The B script looked neater and more stylized than the A. Most Linear A texts were incised directly onto unruled clay, giving the writing a somewhat scattershot appearance. In contrast, Linear B tablets were nearly always ruled: The text sat on tidy horizontal lines that had been cut into the wet clay before the writing began. "Evidently the tablets were supplied in this state to the clerk, like ruled sheets of paper in a modern business office," Evans wrote.

Of the three Cretan scripts Evans discovered—hieroglyphic, Linear A, and Linear B—Linear B stood the best chance of being deciphered. As with any secret code, the more text a decipherer has to work with, the greater the likelihood of solution. The number of Linear B tablets unearthed at Knossos far outstripped any other kind; over time, more than two thousand would be found there. From the beginning, it was Linear Script Class B, used in the twilight days of the Palace of Minos, that held the greatest promise.

The Linear B tablets have a stark beauty. Some have smooth, charcoal-gray surfaces resembling slate, others are reddish brown, still others are bright orange. (The color depends on the level of oxygen to which they were exposed when the palace burned down.) The incised characters are generally crisp and made with care. They are, as Evans put it, "the work of practised scribes." On the backs of tablets, those scribes left traces of themselves in the form of fingerprints and even doodles. To look at the tablets even now is to be in the presence of other people—living, thinking, literate people.

This feeling animates all archaeological decipherment. The pull of an undeciphered ancient script comes not only from the fact that its discoverer cannot read it, but also from the knowledge that once, long ago, *someone could.* To Evans, the scribes of Knossos were real people who had set down the workings of their Bronze Age world, precisely and deliberately, on pieces of wet clay. Men could read those tablets once. It should be possible, even after thirty centuries, for man to read them again.

Three-thousand-year-old scribal doodles from Knossos (*top*) and the Greek mainland (*bottom*).

"We have here locked up for us materials which may some day enlarge the bounds of history," Evans's assistant John L. Myres wrote in 1901. "The problems attaching to the decipherment of these clay records are of enthralling interest."

And so they would be, for fifty years to come.

2

THE VANISHED KEY

ABOUT FIVE THOUSAND YEARS AGO, man stopped having to remember quite so much. Spoken language had already been in existence for at least fifty thousand years, evolving hot on the heels, evolutionarily speaking, of the dawn of man. But it wasn't until long afterward that man realized he could set down his language in graphic form, using visual symbols to encode speech and store it for later retrieval. For the first time, people did not have to depend on memory alone to transmit the history, lore, and daily activities of their communities. We call these marvelous storage-and-retrieval systems writing.

One of the foremost inventions in the history of mankind, writing probably developed independently in several places at around the same time. Before that time, people relied on a range of crude systems, like knotted strings, clay tokens, or notches cut in sticks, to help them count, tally, and remember. Scholars call these proto-writing. But writing proper—a full symbolic system that can record any imaginable text in a community's language—began only with the advent of Sumerian cuneiform in about 3300 B.C. In a separate though possibly related development, the hieroglyphs of ancient Egypt arose around the same time.

Writing systems seem to have been rare in the ancient world, and even today, they are something of a linguistic luxury: By some estimates, only about 15 percent of the roughly six thousand languages spoken around the globe have written forms. It is entirely possible to have language without writing. Not so the other way round.

A writing system is simply a map. It works by taking the sounds of a language and mapping them, singly or in combination, onto designated graphic symbols. There are three types of mapping possible—three ways, that is, in which the sounds of language can be made to meet the eye. Every writing system in the world is one of these three types, or some combination of them.

The first type, in which a written symbol stands for a whole word (or a whole concept), is called logographic (or ideographic) writing. Chinese writing, with its tens of thousands of characters, each signifying a different word of the language, is the best-known example of a logographic script.

In the second type, a symbol stands for a single syllable, like *ma* or *pa, bo* or *do, tam* or *kam.* Examples of syllabic scripts, or syllabaries, as they are also called, include the Japanese kana script. (Japanese writing as a whole is a mixed script; besides syllabic characters, it includes a great many logograms, which like the kana are borrowed from Chinese script.)

In the third type of writing system, symbols stand for individual sounds. This is an alphabet. We owe the alphabet to the Phoenicians, a Semitic people, who fashioned a letter-by-letter writing system from an earlier Semitic script in about 1000 B.C.

As the Phoenicians conceived it, the alphabet had twenty-two characters—consonants only, no vowels. This alphabet was later taken up by the Greeks, who added vowel signs; from the Greeks it passed to the Etruscans and on to the Romans, who gave us the familiar alphabet used to write many Western languages, including English. The Phoenician script and its immediate descendants are the progenitors of nearly all the alphabets used round the world today, from Roman to Cyrillic to Hebrew and Arabic and, quite probably, many of the graceful curvilinear scripts of India.

Whichever type it is, every writing system operates on the same basic principle, using individual symbols to represent one or more sounds of a language. In logographic systems, a character stands for the whole string of sounds known as a word. A syllabic system bites off smaller chunks, using a character to stand for perhaps two or three sounds. In an alphabetic system, the segment is smaller still, usually just a single sound, as with our *l* or *m* or *t* or *v*. In this way, every writing system functions as a kind of elementary encoding device, mediating between spoken sound and graphic symbol. But in order for the code to work properly, it must be transparent to all who would use it.

To know the relationship between the sounds of a language and the written symbols that represent them—to hold the key to the code—is to be able to read that language. As long as there is someone alive who retains the key, the language can be read. But with the passage of time, the key can be lost. Now the link between sound and symbol is broken, and the text becomes as impenetrable as any secret code. That is where decipherment comes in.

In his book *The Story of Archaeological Decipherment,* the scholar Maurice Pope elegantly describes the lure of ancient script:

Decipherments are by far the most glamorous achievements of scholarship. There is a touch of magic about unknown writing, especially when it comes from the remote past, and a corresponding glory is bound to attach itself to the person who first solves its mystery. Moreover a decipherment is not just a mystery solved. It is also a key to further knowledge, opening a treasure-vault of history through which for countless centuries no human mind has wandered. Finally, it may be a dramatic personal triumph.

When a reader confronts a text, the crucial question is this: What is known about the language of the text and what, correspondingly, is known about the script used to write it? For any written text, this relationship between language and script can take just one of four forms. Diagrammed, they make a tidy four-cell table:

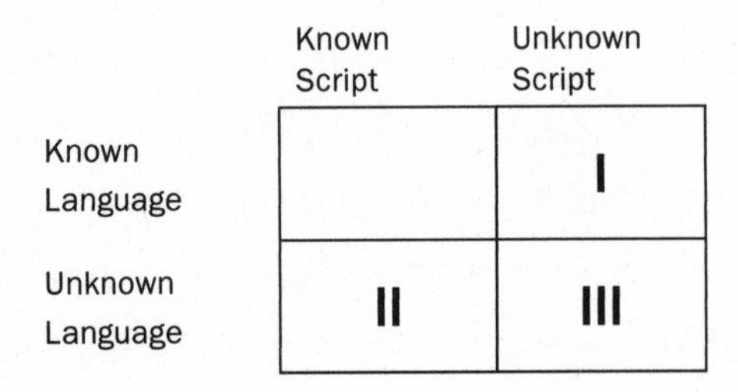

The upper-left-hand cell, which is blank, represents the most straightforward case. Here, a known language is written with a known script, as in the passage you are now reading, written in English by means of the Roman alphabet. No decipherment is required.

The other three cases entail decipherments of varying difficulty. In Case I, in the upper-right-hand corner, an unknown script is used to write a known language. That is the case for Rongorongo, an undeciphered script of Easter Island discovered in the 1860s. Dating to the eighteenth century A.D. or earlier, the script was apparently used to write Rapa Nui, a Polynesian language still spoken on the island. But because Rongorongo fell into disuse (and because it lacks many of the internal clues, like word breaks, that help analysts tease a script apart), it is now impossible to tell which symbols in the script correspond to which sounds of the language.

In Case II, on the lower left of the table, a known script is used to write an unknown language. That happened with Etruscan, an ancient non-Indo-European language of Italy that remains poorly understood to this day. The script used to *write* Etruscan is known: It derived from the Greek alphabet. As a result, it is still possible to read an Etruscan text aloud, giving each letter its familiar Greek sound-value. But to even the most knowledgeable linguist, the result sounds like gibberish. No one knows what most Etruscan words mean, or how Etruscan grammar worked. Read aloud, an Etruscan text is awash in sound but signifies practically nothing.

Case III, on the lower right, is the toughest of all. Here, both language and script are unknown. This is the most inhospitable environment for decipherment possible, for it gives the decipherer no outside aid: no familiar script to help sound out the language, no familiar language to help sort the script. Such was the case for Linear B when Evans unearthed it in 1900. The script was a linguistic terra incognita with neither map nor compass at hand.

* * *

IN EVANS'S DAY, the most famous archaeological decipherment in history was that of the Egyptian hieroglyphics. Unlike Linear B, the hieroglyphs of ancient Egypt did not have to be dug up and discovered: They were always there, and they always beguiled. Their decipherment, accomplished in the 1820s, would exert a considerable and ultimately harmful influence on Evans's approach to Linear B.

Developed in about 3000 B.C., Egyptian hieroglyphic writing was in use for more than three millennia. After that, with the spread of Christianity, the Egyptian language was written increasingly with the Coptic alphabet, a twenty-four-character script derived from the Greek alphabet. By about 400 A.D., the hieroglyphs had been abandoned entirely. Carved in stone, they would tantalize scholars for centuries to come.

By the modern era, no one was even certain what language the ancient Egyptians had spoken. Whatever it was—possibly an ancestor of Coptic, an Afro-Asiatic language later used in the region—it had long since been supplanted by Arabic. Faced with the decipherer's worst-case scenario, an unknown script writing an unknown language, generations of scholars could do little more than speculate on what the hieroglyphs said.

In 1799, a key appeared at last with the discovery of the Rosetta stone. The Napoleonic Wars were raging, and one of Napoleon's campaigns had brought a troop of French soldiers to the Egyptian town of Rashid, known in the West as Rosetta. The soldiers were charged with dismantling an ancient wall there. As they did, they came upon a large black slab set into the wall. Weighing three-quarters of a ton, it was removed to Cairo for study and eventually made its way to the British Museum, where it resides today.

There were three scripts on the stone. On the bottom was a passage in Greek, which could be read with ease. It was a decree from 196 B.C. describing, as the journalist Simon Singh wrote in *The Code Book,* "the benefits that the Pharaoh Ptolemy had bestowed upon the people of Egypt, and . . . the honors that the [Egyptian] priests had, in return, piled upon the pharaoh."

The other two scripts wrote the ancient Egyptian language. On top was a passage in the familiar ornate hieroglyphs. In the middle was a passage in the style of Egyptian writing known as demotic. Cursive, streamlined, and faster to write than hieroglyphics, the demotic script had been introduced in the seventh century B.C. Where the Egyptians had used hieroglyphs for religious, dedicatory, and other official inscriptions, they used demotic for everyday writing.

Two languages, three scripts. In the Rosetta stone, prospective decipherers had found their hoped-for bilingual inscription—known in archaeologists' parlance simply as "a bilingual." Since the stone was obviously meant as an official record, it was reasonable to assume that all three passages said the same thing.

Thanks to the Greek passage, the general meaning of the hieroglyphic text was known. But the precise way in which the hieroglyphics encoded the Egyptian language—what type of writing system they were and what the sound-value of each character was—remained a mystery. Because of the highly pictorial nature of the glyphs, most scholars assumed they were part of a logographic system, with each little picture representing a separate Egyptian word.

A partial answer came in 1814 from Thomas Young, an English physician and polymath. Born in 1773, Young was known for conducting important research on color perception and the physiology of the human eye and, in particular, for his work in physics. He was also deeply interested in languages. "Young was able to read fluently at the age of two," Singh wrote. "By the age of fourteen he had studied Greek, Latin, French, Italian, Hebrew, Chaldean, Syriac, Samaritan, Arabic, Persian, Turkish and Ethiopic, and when he became a student at Emmanuel College, Cambridge, his brilliance gained him the sobriquet 'Phenomenon Young.' "

Poring over copies of the Rosetta inscriptions, Young began to suspect that the Egyptian hieroglyphs were not a pure logographic system after all, but rather a mixed script, comprising both pictograms and phonetic characters. He concluded this after he matched some Greek proper names on the stone to whole strings of hieroglyphics rather than to single signs.

Proper names are a decipherer's best friend. They normally pass from one language to another almost unchanged, and it is often possible to pick them out precisely, even in an unfamiliar script. (They would ultimately play a significant role in the decipherment of Linear B.) Seeing a proper name on the known side of a bilingual immediately gives the decipherer something to look for on the unknown side. Suppose, for instance, you unearth this tiny tablet:

Τῷ Ὅμηρῳ χρῦσὸν Μαργαλίτις πέμψει.

Margalit will send gold to Homer.

Inscribed in English and Classical Greek, the tablet contains two proper names. Even if you have never encountered Greek—and even though the words appear in a different order than in the English passage—you can make a pretty fair guess as to which these are.

Young seized on the fact that in Egyptian writing, groups of hieroglyphics were sometimes ringed by little enclosures, known as cartouches. There were a number of cartouches on the Rosetta stone. In 1762, Jean-Jacques Barthélemy, a French priest who was a scholar of Eastern languages, had made the inspired guess that the cartouches set off words of great importance, such as the names of gods or rulers. One of the names in the stone's Greek text was *Ptolemaios,* the pharaoh Ptolemy. If Young were able to locate the name Ptolemy in the Egyptian text, Singh wrote, "it would enable [him] to discover the phonetics of the corresponding hieroglyphs, because a pharaoh's name would be pronounced roughly the same regardless of the language."

Confounding the problem was the fact that the arrangement of symbols in a cartouche was rarely fixed. Though the overall direction of Egyptian script was right to left, within a cartouche, scribes tended to group characters in whatever configuration was most aesthetically pleasing. Despite these difficulties, one cartouche seemed especially promising. With certain variations, the scribes had repeated it on the stone a half-dozen times. In its simplest form, it looked like this:

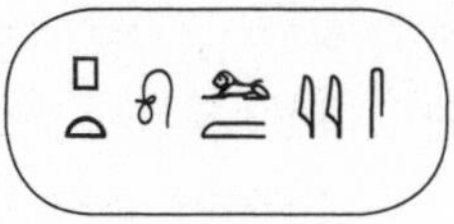

Could this cartouche spell the name Ptolemy? By trial and error, Young began assigning phonetic values—that is, *sound-values*—to the seven different glyphs inside it. "Although he did not know it at the time," Singh wrote, "Young managed to correlate most of the hieroglyphs with their correct sound values." As Young suspected, the characters in the cartouche weren't logograms at all. Instead, they worked phonetically, with each character representing a single sound of Egyptian. Here are the actual sound-values of the glyphs in the Ptolemy cartouche, most correctly determined by Young:

Symbol		Sound-Value
𓊪	=	"p"
𓏏	=	"t"
𓍯	=	"o"
𓃭	=	"l"
𓐝	=	"m"
𓇌	=	"e"
𓋴	=	"s"

Despite his discovery, Young remained under the powerful spell of iconicity. Like most people who sought to decipher the Egyptian hieroglyphs, he could not resist the seduction of their pictorial forms, which resembled tiny flora and fauna, pots, pans, and people. Perhaps, Young reasoned, the Egyptian scribes had reserved phonetic spelling for foreign names, using logograms for everything else. Hampered by this assumption, he could progress no further.

The true decipherment of the hieroglyphs was accomplished by Jean-François Champollion. Born in France in 1790, Champollion was also a boy wonder. When he was still a teenager, after coming across an account of the Rosetta expedition in his brother's scholarly library, he vowed to decipher the hieroglyphic script himself. He set to work, though a successful solution took him until the ripe age of thirty-four.

In his late teens, Champollion wrote a book, *Egypt under the Pharaohs,* eventually published 1814. In it, he argued that the language of ancient Egypt was Coptic itself, preserved in his own time in the liturgy of the Christian Coptic Church. It was an idea that would figure significantly in his decipherment a decade and a half later. At twenty, he began teaching at the University of Grenoble. In the coming years, after having read one or two of Young's articles on the subject, he began to prepare to attack the Egyptian hieroglyphics.

To do so, Singh wrote, he studied "Latin, Greek, Hebrew, Ethiopic, Sanskrit, Zend, Pahlevi, Arabic, Syrian, Chaldean, Persian and Chinese." It was crucial, Champollion knew, to understand the workings of languages from as many different linguistic families as possible. If the language of the hieroglyphs turned out not to be Coptic, the structure of any one of these might shed light on its real identity. (Coptic was no problem in any case: Champollion had already learned the language in his teens. He was so fluent, Singh wrote, "that he used it to record entries in his journal.")

As Champollion studied the hieroglyphic inscriptions, he realized that most names (and not just the foreign names, as Young had supposed) were spelled out phonetically. A lovely example was found in this small cartouche:

𓍹𓇳𓄟𓋴𓋴𓍺

The final glyph, 𓋴𓋴, was already known to stand for "s"—it is simply a double version of the one in the Ptolemy cartouche. The values of the first two glyphs were unknown. Here, Champollion made a remarkable conjecture. The first glyph, 𓇳, looked like a stylized sun. What if this symbol were actually pronounced like the *word* for "sun" in the Egyptian language? The Coptic word "sun" was *ra*. If the inscription were written in Coptic, then the cartouche so far would read Ra-___-s. That left only the second glyph, 𓄟, which other inscriptions suggested stood for the consonant cluster "ms." The word in the cartouche, then, was Ra-ms-s, the native Egyptian name of the great pharaoh Ramses. (In accordance with Egyptian spelling conventions, which did not take great pains to represent vowels, the scribes had left out the "e.")

Despite its highly pictorial appearance, then, Egyptian hieroglyphic writing was a mixed script. Many symbols, like 𓊪, 𓏏, 𓍯, 𓃭, 𓋴, and 𓄟, worked alphabetically, representing single sounds or sound-clusters. Others, like 𓇳, worked on the rebus principle, symbolizing a small string of sounds by functioning as little puns. The puns worked, Champollion demonstrated, only if the language of the inscriptions was Coptic.

Champollion's decipherment brought home a crucial point: In most scripts, the *form* each character takes is completely arbitrary, and any script can employ characters of any form. Though they looked like pictures of real-world objects, some Egyptian hieroglyphs actually stood for *sounds.* Many nineteenth-century analysts of ancient scripts, swept along on a Romantic tide of iconography, never grasped this essential point. Though he ought to have known better, Arthur Evans was one of them.

For Evans, the problem stemmed from the fact that the Egyptian scribes often *did* use hieroglyphs to stand for whole words or concepts. Besides using phonetic characters, Egyptian hieroglyphic script contained a set of pictograms that could be appended to spelled-out words. Known as determinatives, these pictograms conveyed additional information about the word, modifying its meaning, distinguishing among multiple meanings, or identifying the larger category—human, animal, royalty—to which a thing belonged. Egyptian determinatives included 𓀀, "man" or "lord"; 𓁐, "woman" or "lady"; 𓀙, "old man"; 𓀭, "deity"; and 𓊖, "crossed roads" or "settlement."

Determinatives can be found, to a greater or lesser degree, in other writing systems. As Ventris's biographer Andrew Robinson points out in his book *The Story of Writing,* when capital letters are used to signal proper names in English and other languages, they are functioning as determinatives of a sort. Whether you were conscious of it or not, capitals most likely helped you pick out the proper names in the Classical Greek "tablet" on page 48.

And it was the presence of determinatives in Linear B—

or something that looked a great deal like them—that would bring Arthur Evans a world of trouble as he began to sift the strange Bronze Age symbols.

THE SCRIBES AT KNOSSOS never imagined they were writing for posterity. They were simply keeping the records of their own community, in their own language, as any chroniclers would do. But with the passage of millennia, those records had become, quite literally, cryptic.

A decipherer approaches an ancient script much as a cryptanalyst does a secret code. In some respects, the decipherer's task is easier: Unlike codes, real-life writing systems are rarely meant to conceal or deceive. In other ways his task is harder—often much harder. Unlike the cryptanalyst, the decipherer may not know what language is being encoded. When you do the cryptogram in your Sunday paper, you are secure from the start that the solution will be in English. The decipherer of a forgotten script may have no comparable assurance.

But even when the language of a script is unknown, the script itself is filled with internal clues, if only one knows where to look. Confronted with a piece of unknown writing, the decipherer must begin by subjecting it to a series of diagnostic tests. Each test is a small forensic exercise designed to coax the script, bit by bit, to yield up its identity.

The first test is so obvious it is sometimes overlooked. The decipherer has to establish at the start whether the tangle of symbols before him actually is writing. This is not always as straightforward as it seems. Suppose you unearth another tablet, one that looks like this:

Does this tablet, with its rows of primitive symbols, contain writing? No: It is a Modernist painting, *Pastorale (Rhythms),* done in 1927 by the artist Paul Klee.

What about these whimsical sculptures? Could they conceivably be writing?

Yes: The photograph shows a fragment of an ancient Mayan tablet. The carvings are Mayan hieroglyphics, a writing system deciphered only in the late twentieth century.

With some symbol systems, the answer is far less clear. Scholars have practically come to blows over Rongorongo, the hieroglyphic script of Easter Island. Carved on wooden tablets, Rongorongo contains hundreds of logograms. In their vibrant stylization, they evoke the stick figures of Sir Arthur Conan Doyle's "Dancing Men" cipher as reimagined by the pop artist Keith Haring:

Detail of a Rongorongo tablet from Easter Island.

Though Rongorongo has resisted decipherment for well over a century, some modern scholars insist it is a true writing system. Others call it a kind of proto-writing, used as a memory aid for traditional rituals performed aloud. To still others, it is visual art and nothing more.

Fortunately for Arthur Evans, Linear B passed this first test handily. Stored in archival boxes and carefully labeled, the tablets were unmistakably documentary records. It was plain to him from the moment he unearthed them that the symbols they contained were writing, meant to be read.

Once the decipherer is sure he has writing, he must establish precisely what kind of writing system he has. In principle, this can be done with a very simple test: Count the number of different characters the writing system contains, and the total will tell you what kind of system it is. If the number is very large—say, in the thousands—you are looking at a logographic system, like Chinese, in which each word of the language is written with a separate symbol. If the number is between about 80 and 200, then you have a syllabic system. The Cherokee syllabary, invented by Chief Sequoyah in 1819 and containing 85 characters, is just such a system:

Ꭰ = a	Ꭱ = e	Ꭲ = i	Ꭳ = o	Ꭴ = u	Ꭵ = v
Ꭶ = ga Ꭷ = ka	Ꭸ = ge	Ꭹ = gi	Ꭺ = go	Ꭻ = gu	Ꭼ = gv
Ꭽ = ha	Ꭾ = he	Ꭿ = hi	Ꮀ = ho	Ꮁ = hu	Ꮂ = hv
Ꮃ = la	Ꮄ = le	Ꮅ = li	Ꮆ = lo	Ꮇ = lu	Ꮈ = lv
Ꮉ = ma	Ꮊ = me	Ꮋ = mi	Ꮌ = mo	Ꮍ = mu	
Ꮎ = na Ꮏ = hna Ꮐ = nah	Ꮑ = ne	Ꮒ = ni	Ꮓ = no	Ꮔ = nu	Ꮕ = nv
Ꮖ = qua	Ꮗ = que	Ꮘ = qui	Ꮙ = quo	Ꮚ = quu	Ꮛ = quv
Ꮝ = s Ꮜ = sa	Ꮞ = se	Ꮟ = si	Ꮠ = so	Ꮡ = su	Ꮢ = sv
Ꮣ = da Ꮤ = ta	Ꮥ = de Ꮦ = te	Ꮧ = di Ꮨ = ti	Ꮩ = do	Ꮪ = du	Ꮫ = dv

Ꮬ = dla Ꮭ = tla	Ꮮ = tle	Ꮯ = tli	Ꮰ = tlo	Ꮱ = tlu	Ꮲ = tlv
Ꮳ = tsa	Ꮴ = tse	Ꮵ = tsi	Ꮶ = tso	Ꮷ = tsu	Ꮸ = tsv
Ꮹ = wa	Ꮺ = we	Ꮻ = wi	Ꮼ = wo	Ꮽ = wu	Ꮾ = wv
Ꮿ = ya	Ᏸ = ye	Ᏹ = yi	Ᏺ = yo	Ᏻ = yu	Ᏼ = yv

The Cherokee syllabary.

If the character count is only in the dozens, then you are looking at an alphabet. The alphabets of the world range in size from about a dozen letters, used by the Rotokas alphabet of the Solomon Islands, through the thirty-three Cyrillic letters used in Russian, to the more than seventy characters of the Khmer alphabet of Cambodia.

With a known script, counting characters is easy. With an unknown script, it can be a nightmare that drags on for decades. Consider the following scenario: One day, an alien lands in Times Square. He is a very bright alien, but he knows no Earth language, nor has he encountered writing of any kind. Assaulted by a blizzard of text from the billboards and newsstands and neon signs around him, he tries to make sense of the Roman alphabet by scrutinizing all its available forms. He sees it on video screens and the printed page, on vertical signs and horizontal ones, in an array of fonts, colors, and sizes. It will take our alien years of minute comparison before he can be positive, say, that 𝒪, 𝟎, and ❂ are mere variations of the same letter, while O, Q, and 𝒞 are all different letters entirely. And so on, for every symbol he sees. There is difference, and there is meaningful difference. You cannot count what you cannot distinguish.

The decipherers of Linear B found themselves in similar straits. As David Kahn wrote evocatively in *The Codebreakers*:

> The individual signs of Linear B are rather fanciful and resemble a whole variety of objects—a Gothic arch enclosing a vertical line, a ladder, a heart with a stem running through it, a bent trident with a barb, a three-legged dinosaur looking behind him, an A with an extra horizontal bar through it, a backward S, a tall beer glass . . . with a bow tied on its rim; dozens look like nothing at all in this world.

After decades of digging at Knossos, Evans had thousands of tablets at his disposal. Their combined texts ran to tens of thousands of signs. He noticed immediately that many signs looked similar though not quite identical, as in these pairs:

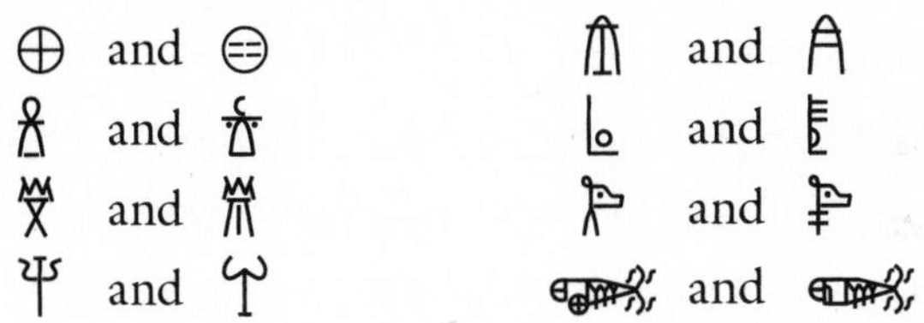

Did each pair show variants of the same sign, or different signs entirely? There was no swift way of telling. One of the American investigators, the Brooklyn College classics professor Alice Kober, spent years agonizing over the symbol , and whether it was the same or different when it had just one horizontal crossbar instead of two, as the scribes sometimes wrote it.

Nor were the Linear B decipherers working from nice, neat typefaces, either. (Today, there are several digital fonts available for Linear B; the one used in this book is known as Aegean.)

Clay is splendid for stamping, as the Sumerian cuneiform tablets plainly show. It is a less good medium for inscription. Wet clay exerts drag, and as a stylus moves through it, it leaves burrs in its wake, distorting the look of the characters. If you are at all doubtful, try writing your name in a block of Plasticine with a straight pin taped to a stick.

There were perhaps seventy scribes at Knossos, and, as with any group of writers, some simply had better penmanship than others. To make matters worse, whenever the Cretan scribes made errors—provided they caught them in time, for dried clay is the original read-only medium—they "erased" them by smearing the clay with a finger (or the flat edge of a wooden stylus) and overwriting the correct sign. This often rendered the new sign blurry and, to modern eyes, hard to interpret.

Evans, at least, had the advantage of being able to see the Knossos tablets firsthand and make careful drawings of the characters. Not so for the other scholars, who had to make do with the few, poor-quality photographs Evans made available. In the end, the task of compiling a definitive list of Linear B signs was so difficult that despite years of work by scholars on both sides of the Atlantic, it was not completed until 1951.

At the same time the decipherer is counting the signs, he must also determine the direction in which the script is written. Most Western scripts run from left to right. Other scripts, like Phoenician, Hebrew, and Arabic, flow from right to left. Still others, like Chinese and Japanese, are written from top to bottom.

Scripts can run in stranger ways. In its earliest days, the Greek alphabet, like its mother, the Phoenician, was written right to left. Eventually, it settled into its familiar left-to-right order—an order that the alphabets descended from it share today. But for a time in between, Greek writing literally flopped back and forth within a single document, running right to left and left to right in alternate lines. The result, a curving thread of script that calls to mind a furrowed field, is known as boustrophedon, Greek for "as the ox turns." The following inscription, written in a version of the Greek alphabet used in the sixth century B.C., is an example of boustrophedon writing. Notice, too, that in the "backward" lines, the direction of each letter is also reversed:

← letters reversed
← letters reversed
← letters reversed

A twenty-first-century rendering of lines from Homer's *Iliad*, written as boustrophedon by Professor Thomas G. Palaima of the University of Texas, in the version of the Greek alphabet used on the island of Euboea in the sixth century B.C. The Euboeans were the great colonizing power of the period, and they carried this version of the alphabet to Italy, where it was adopted first by the Etruscans and later by the Romans, from whom our present-day Roman alphabet is descended.

Even odder-looking is reverse boustrophedon, in which every other line is written upside-down. (The Germans call this style *Schlangenschrift*, or "snake writing.") Rongorongo is written like this, suggesting that the reader had to turn the tablet 180 degrees at the start of each line.

There are several ways to determine the direction of an unknown script. Punctuation helps, of course, but not all writing systems use punctuation. The most effective test is to scan the writing surface for "white space"—places where the text has filled out a line incompletely. If our alien were studying this book, the unfilled spaces at the right-hand ends of many paragraphs would strongly suggest that the Roman alphabet runs left to right. Knowing the script direction gives him essential information about English spelling: He now knows, for example, that the word just before the colon in this sentence is *spelling* and not *gnilleps*. Such distinctions are crucial to decipherment.

As Evans examined the tablets, he made use of similar clues. On large tablets, long texts were broken up into paragraphs, which often ended with empty space on the right-hand side. He knew then that Linear B ran from left to right.

The decipherer's next task is to find word breaks. Many writing systems mark divisions between words with spaces or by other means. But some systems do not. Boustrophedon writing, as in the early Greek example above, arranges characters in one long, sinuous line, with no spaces between words.

A script with word divisions gives the decipherer a running start. Imagine trying to solve a cryptogram in which all the characters are run together, like this:

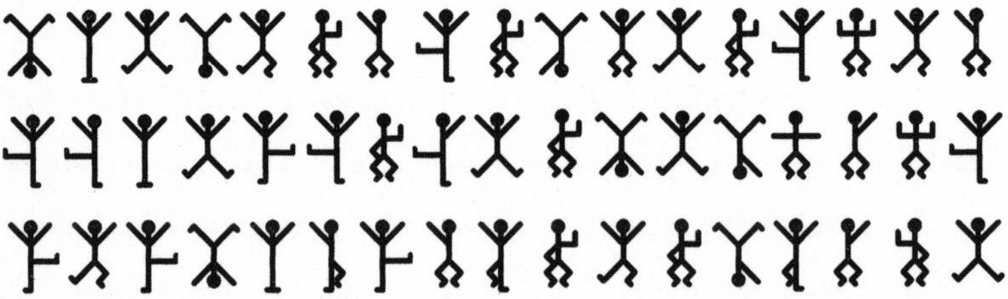

The script is a modern version of the Dancing Men cipher, invented by Sir Arthur Conan Doyle for "The Adventure of the Dancing Men," a Sherlock Holmes story from 1903. The secret messages in Conan Doyle's original story required only seventeen discrete characters; the designers of the several Dancing Men fonts available today have created the rest.

As it appears above, the cipher is one continuous stream, a mass of black symbols with no breaks of any kind. If this were an ancient script, its decipherer would be at a deep disadvantage: Without knowing where words start and end, he has no ready way to tell whether the system is logographic, syllabic, or alphabetic, much less what the language might be—or even if the strange symbols are writing at all.

Now look at the same text, this time written with the symbols as Conan Doyle meant them to be used:

At first glance, the two blocks of text look alike. But closer inspection reveals that some figures in the second cipher are holding tiny flags. In the cipher on which "The Adventure of the Dancing Men" hinges, Conan Doyle used these flag-bearing figures to signal word breaks. Holmes, being Holmes, picked up on the flags immediately and cracked the code.

With the word breaks marked, the second text gives a decipherer far more to work with. He now knows that the cipher contains a one-letter word (the character), which, assuming the text is in English, is almost certainly "a" or "I." The text also contains a three-letter word, , which is a good candidate for "the." (Both texts above read, "The Dancing Men is a cipher invented by Sir Arthur Conan Doyle.")

In this respect, too, Evans was lucky. On the tablets, the Linear B scribes had thoughtfully separated groups of signs with small vertical tick marks, as in this six-word line, part of a longer inscription:

Each sign-group, Evans knew, was almost certainly a word of the Cretan language.

BY THE END of the second millennium B.C., there was no one left who held the key to Linear B. There is historical evidence that long before Evans dug them up, the tablets had already begun to tantalize. In A.D. 66, during the reign of the Roman emperor Nero, a violent earthquake at Knossos flung to the surface what appeared to be a tin chest. The shepherds who found the chest looked eagerly inside for treasure but discovered, Evans wrote, only "documents of 'lime-bark,' inscribed with 'unintelligible letters.' " The documents were brought to the emperor, who, supposing them to be Phoenician, summoned Semitic-language experts to interpret them. The experts were doubtless under pressure—pressure, most likely, of being allowed to continue breathing—to give the emperor the answer he wanted. After studying the tablets, they determined that they were indeed written in Phoenician and represented, as Evans wrote, "the journal of one of the ancients—the Knossian Diktys, companion of Idomeneus, who had been present at the Trojan War."

In fact, Evans wrote, these ancient "lime-bark" records might well have been clay tablets from the Knossos palace, thrust aboveground by an earthquake more than a millennium after they were written. The Knossos scribes were known to have stored some of their file boxes in hollowed-out stone chests, known as cists, which were lined with lead sheeting to protect the contents. While the wooden file box within would long since have decayed, the lead-walled cist itself might well be mistaken for a tin chest. Furthermore, he wrote, "The brown, half-burnt tablets of the Palace themselves bear a distinct resemblance to old or rotten wood"—hence the reference to "lime-bark" in the Classical sources describing the find.

The curious documents Emperor Nero held in his hands may have been the records of the Palace of Knossos, written in Linear B by court scribes two centuries before the Trojan War. Nero was reported to have ordered the documents translated from "Phoenician" into Greek and to have filed the translation away in his personal library. Then the tablets were forgotten again, for more than eighteen centuries.

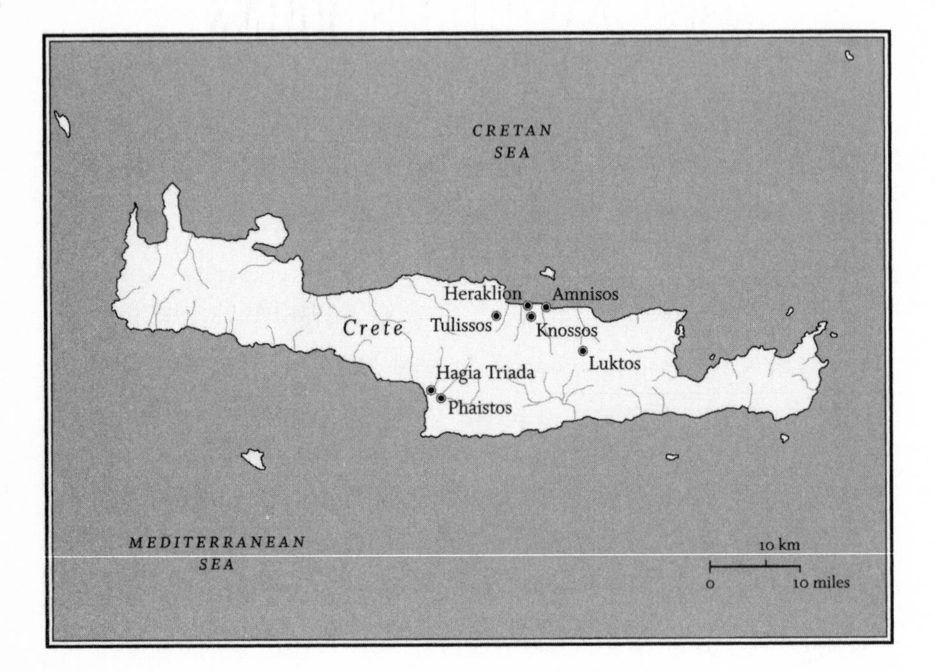
CRETAN
SEA
Heraklion
Amnisos
Tulissos
Crete
Knossos
Luktos
Hagia Triada
Phaistos
MEDITERRANEAN
SEA
10 km
0
10 miles

3

LOVE AMONG THE RUINS

IT TOOK FIRE TO GIVE us Linear B. In about 1400 B.C., the final conflagration at Knossos destroyed most of the palace and its contents, marking the end of the great civilization that had been rooted there for centuries. But the blaze had one completely beneficial effect: It preserved for future generations the clay tablets that recorded the palace's final year.

Cretan scribes never fired their work. In a warm climate like the Aegean there was no need: Inscribed, the damp clay was simply left to dry in the sun, and that was usually enough. Evans discovered this practice to his dismay after he had painstakingly dug up a cluster of tablets from a less fire-damaged area of the palace. Carrying them back to his rented Cretan house for safekeeping, he unwittingly placed them under a rotten spot in the thatched roof. That night the rains came, and he awoke in the morning to find the precious unbaked records reduced to mud.

In other parts of the palace, however, where the flames had burned hotter, tablets were baked to a permanent hardness. "In this way fire—so fatal elsewhere to historic libraries!—has acted as a preservative of these earlier records," Evans wrote after the first season's dig.

But baking also made the tablets dry and brittle, and whenever Evans unearthed a cache of inscribed clay shards, his first job was to fit them back together again like the pieces of a jigsaw puzzle. Often, fragments of a single tablet were widely separated, scattered by animals and earthquakes. (Parts of some tablets were so far-flung they would not be reunited for decades.) Making the job even harder was the fact that pieces of a single tablet might bake at different temperatures—some wound up closer to the fire's center, others farther away. When this happened, they shrank at different rates. The result was a tablet whose disjointed parts no longer looked as though they had ever fit together.

Once Evans had enough complete tablets to work with, he could start to compare the two linear scripts, A and B. He had unearthed just a few Linear A tablets at Knossos. (As it turned out, they would be the only significant examples of the script ever found there.) But in 1902, a colleague, the Italian archaeologist Federico Halbherr, uncovered a cache of Linear A tablets at Hagia Triada, a site in southern Crete. With these, Evans had a sufficient sample for comparison.

The A and B scripts had in common more than fifty characters that were identical or nearly so. Both scripts had been used by Cretan scribes to record commodities vital to the local economy: grain, oil, livestock, wine, and the like. (There were some additional Linear A inscriptions, found on ceremonial artifacts, that contained what appeared to be religious dedications.) In both scripts, text was written from left to right. Where Linear B almost always indicated word breaks (usually by means of the vertical tick marks), Linear A did so only sometimes, using a variety of devices, like little dots between groups of signs. Unlike the Linear B tablets, Linear A tablets were nearly always unruled, and in general the A script looked cruder. It was quite reasonable to assume, as Evans did, that the B script was a later, more refined outgrowth of the A, and indeed this turned out to be the case.

Before the decipherment of either script could begin, Evans needed to answer an essential question: Did they record the same language? He thought they did—they were too similar-looking for it to be otherwise. "The conclusion has been drawn," he wrote, "that the language itself was practically identical and that the differences visible in B must be rather due to dynastic than to racial causes."

Besides sharing individual characters, the two scripts shared a means of writing numbers. In unearthing the Knossos palace, Evans had uncovered the first European bureaucracy, and the tablets, he knew, were the palace's account books. With civilization comes stuff, and with stuff comes the need to keep track of it. Not surprisingly, the palace scribes were great enumerators. The tablets filed neatly away by subject counted everything in the Cretan kingdom from sheep, horses, and swine to footstools, bathtubs, chariot wheels (intact), and chariot wheels (broken). Other tablets contained what appeared to Evans to be census data about the kingdom's human inhabitants—who ran the gamut from monarchs to slaves—including their tax records. As a result, Linear B tablets teemed with numbers.

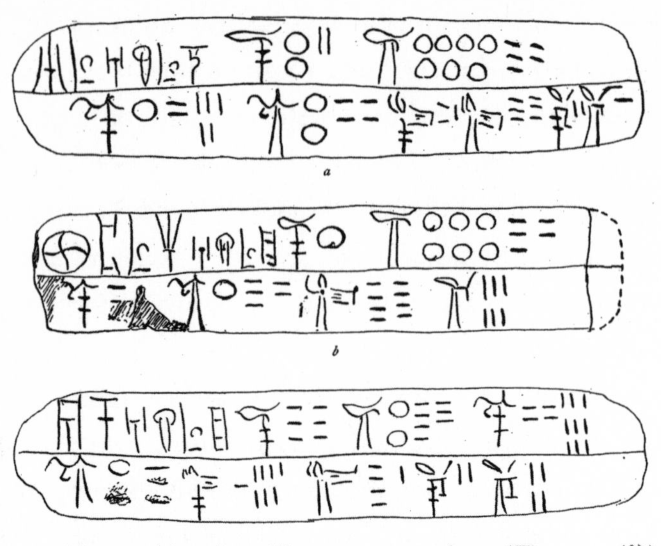

These tablets from Knossos count sheep (𐂃), goats (𐂅), cattle (𐂉), and swine (𐂇).

Evans was able to work out the numerical system quickly. The Cretans used a base-10 system, like our own decimal system. Unlike ours, it was notated by means of only five symbols:

1	=	𐄇
10	=	𐄐
100	=	𐄙
1,000	=	𐄢
10,000	=	𐄫

To write the number five, for instance, the Cretan scribe repeated the "1" sign five times: 𐄋. To write the number fifty, he drew five "10" signs: 𐄔. Fifty-five was written this way: 𐄔𐄋. And so on, for 555 (𐄝𐄔𐄋); 5,555 (𐄦𐄝𐄔𐄋); and 55,555 (𐄯𐄦𐄝𐄔𐄋). The system went up to 99,999, or 𐄳𐄪𐄡𐄘𐄏. There was no sign for zero.

When he was finished with the numbers, Evans turned to the words. The tablets contained many obvious logograms—pictographic signs standing for whole words. These often appeared next to numbers to show the thing counted. One cache of tablets, which came to be known as the "Armoury Deposit," contained inventories of military equipment such as wheels, chariot frames, and javelins. Elsewhere in the palace, Evans came upon a tablet inscribed with the "arrow" sign, , which tallied large quantities of arrows. Nearby were the remains of a chest with more than eight thousand arrows inside.

There were dozens of logograms in Linear B. To modern eyes, some look highly cryptic, among them , , , , , , , , , , and . The meaning of many of these is still debated today.

Others were far more transparent. Some, though conceived in the Bronze Age, are enduringly recognizable, like and , the Linear B signs for "man" and "woman," which would be at home today on any restroom door in the world. Others, including these, were also easily understood:

"horse"	"deer"
"footstool"	"bathtub"
"chariot"	"wheeled chariot"
"wheel"	"spear"
"sword"	"arrow"
"corselet"	"helmet"
"barley"	"spice"
"month"	"tree"

"wine" (depicting the trellises on which grapes were grown)

Evans also identified pairs of related logograms that appeared to stand for male and female animals. One sign in each pair was written with two crosshatches on its stem, while the other had a V-shaped stem. There were signs for stallions and mares (,); boars and sows (,); bulls and cows (,); rams and ewes (,); and billy goats and nanny goats (,). Evans could only guess which sign indicated the male of the species and which the female; the elegant answer to this little puzzle, which came from Alice Kober, would not be known until decades later.

Linear B was also awash in logograms denoting vessels of all kinds:

More than five decades after Evans first dug at Knossos, a tablet inscribed with pictures of humble pots like these would unlock the mystery of Linear B once and for all.

BESIDES LOGOGRAMS, LINEAR B contained dozens of nonpictographic signs. These appeared most often in small strings, called "sign-groups," with the little word-dividers separating each group. These included the ones David Kahn described so evocatively in *The Codebreakers*: the "Gothic arch enclosing a vertical line" ( ), the "heart with a stem running through it" ( ), the "three-legged dinosaur looking behind him" ( ), the "backward S" ( ) and the "tall beer glass . . . with a bow tied on its rim" ( ).

Then there were the symbols that looked, as Kahn wrote, "like nothing at all in this world." Among them were these:

By Evans's initial count, there were at least eighty such characters, a figure that strongly suggested the presence of a syllabary. A further clue came from the number of characters in each sign-group: there were normally between two and five, which also pointed to a syllabic system. As the anthropologist E. J. W. Barber explains in her book *Archaeological Decipherment,* counting the signs in each group is another fine diagnostic tool in the decipherer's arsenal:

> The number of signs between word boundaries will give another indication of the type of script. The words of very few languages consistently run to more than five syllables. . . . Consequently if the number of signs between word breaks is usually around two or three, the script is probably syllabic, whereas if the number is more often eight, ten, twelve, or higher, it is most likely alphabetic.

Linear B appeared, then, to be a mixed script, part syllabic, part logographic. (In this respect, it is not unlike modern Japanese writing.) Evans himself suspected as much: The trick of counting characters to pinpoint the type of script was already well known in his day. But though he knew Linear B was largely syllabic, the presence of so many pictographic symbols would ultimately seduce him, severely hampering his efforts to decipher it. As a result, his assault on the script would occupy him for much of the next forty years.

DURING THESE YEARS, scholars around the world were clamoring to see the tablets. "No effort will be spared to publish the whole collected material at the earliest possible moment," Evans had proclaimed in print after the first season's dig. But when, in 1909, he published *Scripta Minoa,* his three-hundred-page book on the Cretan scripts, only about two dozen pages were devoted to Linear A and B combined. The rest of the text was given over to Evans's painstaking description of the earlier, hieroglyphic script of Crete. (While the book expounded on the symbolism of individual hieroglyphic characters, the author did not attempt to translate the hieroglyphic inscriptions themselves, a feat he wisely deemed impossible.) Although Evans promised additional volumes giving full accounts of the linear scripts, none materialized in his lifetime. As generations of would-be decipherers chafed bitterly, the Knossos tablets remained locked away.

Evans had his reasons. Under a time-honored anthropological tradition that owes much to the colonial imperative, the first investigator to set foot in a village or excavate a ruin retains an unspoken proprietary interest in the place. Future claimants enter at their peril. But there is also a tacit if unspecified time limit by which the original scholar must publish. If he fails to do so, then his turf is fair game.

As the first decades of the twentieth century ticked away, Evans's time limit appeared to have stretched to unsporting length. Though he gave numerous accounts of his finds to mainstream papers like the *Times* of London, he published little significant analysis of the tablets in scholarly journals. Not only did he decline to make most of the tablets themselves available for study, but he also appeared loath to publish drawings or photographs of them: Of the more than two thousand tablets Evans eventually unearthed at Knossos, he would publish reproductions of fewer than two hundred during his lifetime. (In the 1930s, Johannes Sundwall, a distinguished Finnish scholar, published copies of thirty-eight tablets he had managed to see in a Cretan museum, an act that brought down the wrath of Evans.)

In fairness, Evans had many distractions. He remained keeper of the Ashmolean till 1908, and his duties there took time. At home in England, he was active in a string of professional organizations, serving more or less simultaneously as president of the Hellenic Society, the Society of Antiquaries of London, the Royal Numismatic Society, and the British Association for the Advancement of Science. (For his services to archaeology, Evans was knighted in 1911.) He was also involved in good works, in particular with the local Boy Scout troop, which he gave the run of Youlbury's grounds. Having no children of his own, he took in two wards, first a nephew of Margaret's and, shortly afterward, the son of an Oxfordshire tenant farmer.

With the start of World War I in 1914, digging on Crete

became impossible for the duration. Evans was still deeply involved in Balkan affairs, and even before the war's end in 1918, he was taking an active hand in the negotiations that led to the creation of the Yugoslav state.

He was also building three houses. The first was Youlbury, an ongoing project that eventually comprised some two dozen bedrooms, a sunken Roman bath, orchards, magnificent gardens, and a great deal else. Its vaulted marble entrance hall, which looked like nothing so much as a Beaux Arts savings bank and could have housed one comfortably, had a mosaic floor set in a labyrinth pattern, with a tiled Minotaur at the center. Standing imposingly nearby were two huge replicas, carved in mahogany, of the throne of Minos.

"Evans' friends variously described Youlbury as 'shocking' or 'fantastic,' depending on their tolerance for sheer bulk," his biographer Sylvia Horwitz wrote. "It defied all architectural principles of proportion or uniformity of style. Vast to begin with, it grew monstrously large as it rambled without reason and sprouted additions to accommodate this fantasy or that whim of its owner."

Youlbury.

The second house was on Crete. Completed in 1906, it was built by Christian Doll, one of the architects Evans had engaged to direct the reconstruction of the Palace of Minos. The house, which Evans called Villa Ariadne, let him live on the island in accustomed style. Not for him the local wine, made from fermented raisins, that his workmen heartily drank: The villa's cellar was stocked with French wines and champagne. As news of Evans's exploits traveled round the world in the popular press, Villa Ariadne received a string of distinguished visitors, among them the financier J. P. Morgan and the novelist Edith Wharton. When not entertaining, Evans superintended the excavation as elegantly turned out as he would be in a London gentlemen's club. "On the hottest days of a Cretan summer he never came to the dig in shirtsleeves," Horwitz wrote.

The third house was the Palace of Minos—or, more accurately, the Palace of Many Minoses. ("Minos," as Evans suspected, was most likely a dynastic title like "king" or "pharaoh" rather than a personal name. Down the centuries, a series of Minoses would have held court at Knossos.) Over years of excavation, the palace emerged as a vast, increasingly complex organism. As each section was revealed, Evans gave it a name. Besides the Throne Room, these included the Queen's Megaron, or great hall, with its elaborate bathroom and graceful mural of leaping dolphins; the Domestic Quarter, with artisans' workshops in which traces of the goldsmith, the lapidary, and the ceramicist could still be discerned; and the Grand Staircase, down which, in 1910, a visiting Isadora Duncan danced an impromptu dance to the horror of Evans's straitlaced Scottish assistant, Duncan Mackenzie.

With his team of artists, archaeologists, architects, and laborers, Evans spent decades clearing rubble, shoring up collapsing landings, restoring shattered murals, and rebuilding crumbling walls. Where the Minoans had used wood and stone, much of the restoration used newer materials like reinforced concrete. The work was controversial: As Evans rebuilt rooms and repainted murals, he imposed his vision—by definition speculative—of what the palace must have looked like thirty centuries earlier. Today, Knossos is a bustling tourist attraction, but whether it reflects the genuine Minoan aesthetic or an ardent twentieth-century fantasy is an open question.

THROUGHOUT THE FIRST decades of the century, scholars were following Evans's few publications on Linear B with rapt interest. Like him, they could only speculate on what the ancient language of the tablets might have been. Just one thing seemed certain: It wasn't Greek. Today, we reflexively associate Crete with Greek speech, but the Palace of Minos flourished long before early Hellenic peoples—the first Greek speakers—were known to have reached the island.

There was another reason the tablets couldn't be Greek, and this one came from Evans himself, an edict handed down from Olympus: The language of Knossos wasn't Greek because the people of Knossos were altogether different from those of the mainland, where the first Greeks would later settle. Bronze Age Cretans were not only different from their mainland contemporaries, Evans emphatically declared, but they were also superior to them in every conceivable way.

In his earliest writings on the Cretan scripts, Evans assumed, quite reasonably, that the civilization at Knossos was simply an outpost of the larger Mycenaean one on the mainland. But as treasure after treasure was lifted from the ruins of the Palace of Minos, and as mural after mural was restored to reveal images of bright flora, leaping animals, and handsome men and women clad in beautiful garments, Evans became convinced that the civilization he had unearthed at Knossos was older, higher, and far better than the rude Mycenae of the mainland. Before long—and his growing infatuation is palpable in his writings—Evans had fallen in love with "his" Cretans, whose art and architecture seemed to him vastly more refined than what Schliemann had found at Mycenae.

"As excavation went on, . . . the 'Mycenaean' culture of the Greek mainland ceased to be an adequate standard for Cretan achievements," the archaeologist John L. Myres, Evans's former assistant, later wrote. The civilization at Knossos, Evans soon came to believe, represented a completely separate culture from that of the mainland. He called his island culture Minoan, after its storied ruler. To Evans, the few examples of Linear B script unearthed on the mainland only proved how pervasive Minoan influence had been: Crete was not a colony of the mainland, he argued; if anything, it was the other way around. Before long, as the classicist Thomas G. Palaima said in a 2002 BBC television documentary about the decipherment, Evans had imbued his vision of the Minoan-dominated Aegean "with the grandeur of the British Empire."

In archaeology, Evans's word was law. If Minoan culture was distinct from Mycenaean, then it followed that its language was different, too. Even if Greek-speaking people had entered Mycenae earlier than supposed, the gradual Hellenization of the mainland would have had no bearing on Crete, with its separate language and culture. It was clear—clear to Evans, at least—that the language of Linear B was the indigenous Minoan tongue, whatever it might have been. Through Evans's dominance in the field, his position was soon clear to everyone else, too. The few scholars who dared to question him met with swift and certain professional punishment, and for the first half of the twentieth century, the idea of Minoan supremacy was almost universally accepted.

This did not stop scholars and members of the public from enthusiastic speculation as to what the Minoan language might actually have been. Candidates ranged from the preposterous (Basque) to the plausible (Etruscan, the lost language of Italy).

All these things conspired to hold up the decipherment of Linear B for decades. But there was another thing, which by itself would have trumped any combination of distraction, ego, and ideology: Arthur Evans was simply in over his head, something this tiny colossus bestriding the Aegean Bronze Age was constitutionally loath to admit.

Some of his trouble owed to pure bad luck: Where the decipherers of the Egyptian hieroglyphics had the Rosetta stone to help them crack the code, no bilingual inscription for Linear B was found during Evans's lifetime, nor has one surfaced even now. In addition, though he pored over the tablets for years, Evans wasn't trained in the kind of methodical analysis that helps a cryptographer crack a secret code—the kind that offers the only possible entry into an unknown language in an unknown script. "Evans does not . . . seem to have had any clear plan for the solution of the script," Ventris's later collaborator John Chadwick wrote. "His suggestions were in many cases sound, but they were disjointed observations and he never laid down any methodical procedure."

Evans also succumbed early to the siren song of iconicity, an easy thing to do given Linear B's pictorial nature. Like every archaeologist of his time, he had studied the Egyptian decipherment closely, but the methods that had worked for Champollion would not work for him. For Evans, a major stumbling block was the concept of determinatives, the little icons inside an Egyptian cartouche that signaled the word's membership in a category, like "male," "female," or "royal."

Because Linear B had so many little icons of its own, at least some of them, Evans concluded, must be determinatives. He found the character 𐀃 especially seductive. To Evans, it looked like a throne, viewed from the side, with a scepter protruding. "The throne, 𐀃, is high-backed . . . like that in the ceremonial Palace chamber," he wrote. Because this sign occurred in many Linear B words, Evans concluded that it was used, at least some of the time, "as an ideogram, and with a determinative meaning." Where the character appeared alongside a symbol depicting an object—a chariot, for instance—"it surely indicates a royal owner," Evans wrote. When it was used near an apparent personal name, he said, "its inclusion at any rate suggests a royal lineage." (As it turns out, 𐀃 represents the vowel sound "o" and nothing more, but this would not be known for half a century.)

In Evans's hands, other Linear B signs took on iconic dimensions. Of the character 𐀊, one of the "masons' marks" found on the wall at Knossos and later on many tablets, he wrote, "It is itself apparently the derivative of a façade or porch of a building." He made similar conjectures about dozens of signs, a fanciful approach that seriously impeded his efforts to decipher the script.

From 1900 till his death in 1941, Evans tried repeatedly to induce a decipherment from the teeming mass of symbols, as if he could compel the language of the tablets to reveal itself through his accustomed force of will. This obstinacy made him choose to disregard a crucial clue hidden in this fragmentary tablet:

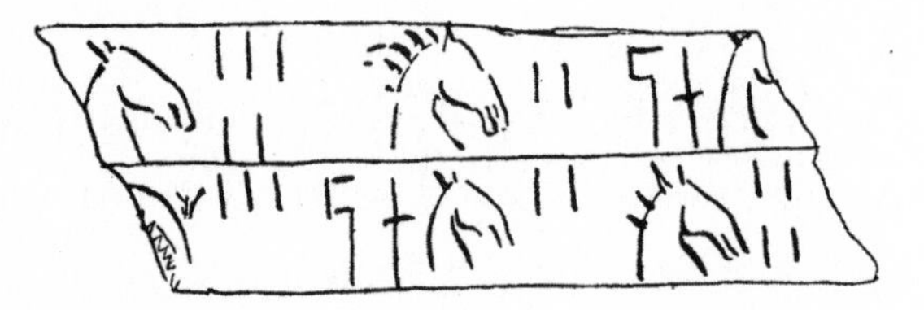

Had Evans allowed himself to follow the clue—deductively—to its logical outcome, he would have been able to unravel Linear B almost immediately.

In the end, as John Myres wrote in Evans's obituary, "the discoverer of the script did not achieve his ardent desire to decipher it." But by the time of Evans's death, that desire was alight in others, including two who would prove far more successful than he.

BOOK TWO

The Detective

Alice Kober in a 1946 photograph from the *Brooklyn Eagle*.

4

AMERICAN CHAMPOLLION

New York, 1946

ON THE EVENING OF JUNE 15, a small, sober-looking woman stood before an audience at Hunter College in Manhattan. The college was her alma mater, and now, two decades after her own luminous career there, she had been invited to address the new crop of Phi Beta Kappa initiates, Hunter's brightest students.

She was by nature self-contained, and speaking in public made her unbearably nervous; each time she did it, she vowed it would be the last. Her talk, like everything else she wrote, was the product of hours of meticulous planning, composition, and revision—she typically put each of her published papers through a good ten drafts until she was satisfied. Before her now was her typescript, its handwritten emendations in her tidy pedagogical script attesting to continued reflection and reworking.

Physically, she was unprepossessing—perhaps, in the parlance of the day, even plain. Short and roundish, she had neat, unstylish hair and dress; solemn, heavy-lidded eyes framed by spectacles; and a thin mouth that gave the impression of primness. Though she was not yet forty, she seemed, like many academic women of her era, prematurely and irretrievably middle-aged.

But as soon as she began to speak, her words had the hypnotic pull of a fairy tale:

> On every kind of writing material known to man, on paper, parchment, papyrus, palm-leaves, on wood, on clay, brick, or stone, on every kind of metal, there exist inscriptions which cannot be read. Sometimes they cannot be read because the system of writing is unknown, and sometimes because, although we know what sounds to ascribe to the different signs, the language is unknown.
>
> These documents range in date all the way from Neolithic times to the present. Some are probably in the process of being written at this very moment. Those, however, that were written at periods which we may call the fringes of history, are especially important for the light that they may cast on the past. . . .

The woman was Alice Kober, an assistant professor of classics at Brooklyn College. By day, she taught a cumbersome load of classes, as many as five at a time—things like Introductory Latin, and Classics in Translation. By night, working almost entirely on her own, as she had for the past fifteen years, she chipped away, methodically and insistently, at the scripts of Minoan Crete. Now, at thirty-nine, although few people knew it, she was the world's foremost expert on Linear B.

Though she is all but forgotten today, Alice Kober singlehandedly brought the decipherment of Linear B closer to fruition than anyone before her. That she very nearly solved the riddle is a testament to the snap and rigor of her mind, the ferocity of her determination, and the unimpeachable rationality of her method. Kober was "the person on whom an astute bettor with full insider information would have placed a wager" to decipher the script, as Thomas Palaima, an authority on ancient Aegean writing, has observed.

Strikingly, she got as far as she did without being able to see any of the tablets firsthand. Even more remarkable was the fact that by the time she addressed the group at Hunter, Kober had already done groundbreaking work on Linear B in an era when there were barely two hundred inscriptions available for study.

That Kober's vital contribution to the decipherment has been largely overlooked is due in great measure to her early death, in 1950, just two years before Michael Ventris cracked the code. It is also due to her quiet, deliberate way of working, step by incremental step, never committing her ideas to print until they met her exacting standards of proof. As a result, she published little—it was more than enough for her, she used to say, to come out with one good article a year. But though her major work spans barely half a decade, from 1945 to 1949, she is now regarded as having built the solid, unassailable foundation on which Ventris's decipherment was erected. For as Ventris himself acknowledged publicly not long before his own untimely death, he had arrived at his solution using the methods Alice Kober had so painstakingly devised.

Along the way, Kober solved a host of small mysteries, many of which had bedeviled investigators from Arthur Evans onward. Among them were these: proving which sign depicted the male animal and which the female in paired logograms like 𐂃 and 𐂂;

correctly identifying the Minoan words for "boy" and "girl," noted together in 1927 and likewise indistinguishable; and being the first to pinpoint the special meaning of the sign ⊜, a discovery previously attributed to Ventris.

But Kober also made much larger discoveries—findings that illuminated the internal workings of Linear B words and symbols—and it was these that had profound implications for Ventris's solution. Her work reads like a how-to manual for archaeological decipherment, something acutely needed for Linear B, a textbook case of an unknown script writing an unknown language. "There is no certain clue to the language of the Minoan scripts," Kober said in a 1948 lecture at Yale. "All we have are the inscriptions they left, and the symbols they contain." She added: "To get further, it is necessary to develop a science of graphics." It was just such a science that Kober, from the first, had set out to construct.

NO ONE BELIEVED Alice Kober when she declared she would make the Minoan scripts her lifework. The year was 1928, and she announced her ambition upon her own graduation from Hunter College. At first glance, she seemed an unlikely candidate to solve a mystery that had already endured for almost three decades. She was young—barely twenty-one—and though she had majored in classics, she had none of the specialized background in historical linguistics that might have put such a calling within reach. Nor was she trained in archaeology, statistics, or any other discipline essential to the decipherer's art.

Above all, she simply did not look the part: With its aura of bravura, derring-do, and more than a dash of imperialism, archaeological decipherment was the time-honored province of moneyed European men. That the upstart American daughter of working-class immigrants would even contemplate the field was dismissed as youthful fantasy.

But in the coming years, on her own time, Kober would systematically acquire every needed weapon in the decipherer's arsenal. She learned a spate of ancient languages and scripts with the methodical ardor of a Champollion. She studied archaeology, linguistics, statistics, and, for their methodology, physics, chemistry, astronomy, and mathematics. All this—more than a decade of concerted study—she did merely to lay the groundwork for her eventual assault on Linear B.

In 1942, Michael Ventris, then only twenty but already passionately interested in the Cretan scripts, wrote confidently from London, "One can remain sure that no Champollion is working quietly in a corner" on the riddle of Linear B. But in fact there was, directly across the Atlantic, sitting quietly at the dining room table of her modest Brooklyn house, ever-present cigarette at hand, and "working hundreds of hours with a slide-rule," as she later wrote. For it was Alice Kober, like Champollion in his day, who imposed scientific precision on the romantic, undisciplined attempts that had gone before. To the riddle of Linear B she brought the skills of a crack forensic analyst in a detective story, who gleans vital information after lesser investigators have trampled through, an unflappable Holmes in a sea of Lestrades. It was only fitting that she, who savored detective stories in what small spare time she had, would give the decipherment the "method and order" she so esteemed.

* * *

ALICE ELIZABETH KOBER was born in Manhattan on December 23, 1906, the elder of two children of Franz and Katharina Kober. Her parents had come to the United States from Hungary earlier that year; Katharina would have been pregnant with Alice when they arrived in May. The couple settled in Yorkville, a historically German and Hungarian neighborhood on Manhattan's Upper East Side. Census records list Franz's occupation as upholsterer and, in later years, apartment-building superintendent. A son, William, was born to the Kobers two years after Alice.

Few traces of Kober's early life are extant. As a teenager, she attended Hunter College High School, one of the city's elite public schools, which, like Hunter College itself, was then for women only. In the summer of 1924, she placed third among 115 New York City high school seniors in a statewide college scholarship contest. Her prize, a hundred dollars a year for four years (about $1,300 a year today), was undoubtedly welcome in a family of modest means. That fall, she entered Hunter College, where she took part in the Classical Club and the German Club.

Even in a college known for its brilliant young women, Kober by all accounts stood out. "As an undergraduate she impressed me by her earnest application to her work, and even more by her independent judgment, which let her accept no statement of the teacher without subjecting it to critical study," one of her professors, the classicist Ernst Reiss, later wrote. "Coupled with this was a still more valuable trait, an intellectual honesty, which induced her readily to revise her own opinion when she became convinced of the correctness of the opposition."

At Hunter, Kober took a course in early Greek life, and it seems to have been there that she encountered the Minoan scripts. In 1928, she was elected to Phi Beta Kappa; she graduated magna cum laude that year, with a major in Latin and a minor in Greek. (C's and D's in gym appear to have put summa cum laude out of reach.) Accepted to graduate school at Columbia, she earned a master's degree in classics in 1929, followed by a Ph.D. in classics there in 1932.

Kober's passage through graduate school, rapid by any standards, was all the more impressive in that she was working the entire time, first at Hunter High School, where she was a substitute Latin teacher, and afterward at Hunter College, where she was an instructor in Greek and Latin. In 1930, two years before she was awarded her doctorate, Alice Kober, then not quite twenty-four, became an instructor at the newly created Brooklyn College at an annual salary of $2,148—just under $30,000 today.

The archaeologist Eva Brann, who studied with Kober at Brooklyn in the late 1940s, recalled her teacher's "dry, refraining rigor" in a biographical essay written in 2005:

> She was, to coin a phrase, aggressively nondescript, or so it seems to me now. She wore drapy, dowdily feminine dresses; something mauve comes before my eyes. Her figure was dumpy with sloping shoulders, her chin heavily determined, her hair styled for minimum maintenance, her eyes behind bottle-bottom glasses snapped impatiently and twinkled not unkindly. Her classroom manner was soberly undramatic, drily down to earth, in no way a performance, but instead demandingly definite. I can't tell whether I re-

> member from observation or infer on reflection that her lectures were very good, forceful and full of matter, but I know that I loved listening.

Toward those she did not respect intellectually (and there were many), Kober could be withering. In a 1947 letter to a colleague, she loosed her scorn on the eminent Czech linguist Bedřich Hrozný. Hrozný had made his name by deciphering ancient Hittite in the 1910s and was just then attempting to do the same for Minoan, with conspicuously less success. "Everybody seems to handle Hrozný with kid gloves," Kober wrote. "I suppose it's because nobody thinks a man with Hrozný's reputation could possibly be as stupid as he seems."

Her harshest criticism, though, was reserved for herself. In 1947, she sent the editor of the journal *Classical Weekly* a lecture on the Cretan scripts he had solicited for publication. Her cover letter read, revealingly: "When you wrote me in May, 1946 I answered that I didn't think it was worth printing. . . . It still seems to me that it isn't quite the thing for the Classical Weekly, though it isn't as bad as I thought it was at the time. . . . The subject is one, as you will see, to which I have a strong emotional reaction, and *emotion and scholarship do not belong together.*"

Her life was her work, and what a great deal of work there was. Kober never married, nor is there evidence she ever had a romantic partner. After her father's death from stomach cancer in 1935, and with her brother grown and gone, she and her mother lived together in the house Alice owned in the Flatbush section of Brooklyn. It was there, night after night, after her classes were taught and her papers graded, that she turned to what she considered the true enterprise of her life: the deep, serious business of deciphering Linear B.

Kober did not write or lecture about the script publicly until the early 1940s, but her private papers make clear that she had begun tackling it long before. "I have been working on the problems presented by the Minoan scripts . . . for about ten years now, and feeling rather lonely," she wrote to Mary Swindler, the editor of the *American Journal of Archaeology,* in January 1941.

Where Evans's approach to Linear B was scattershot, impressionistic, and anecdotal, from the start Kober imposed more rational methods. Her first order of business, starting in the 1930s, was frequency analysis: the creation of statistics "of the kind so successfully used in the deciphering and decoding of secret messages," as she wrote, for every character of the script.

Anyone who has solved a Sunday newspaper cryptogram has met frequency analysis head-on. At its simplest, it entails pure counting, with the decipherer tabulating the number of times a particular character appears in a particular text. If the text is long enough, the frequency count for each letter should mirror its statistical frequency in the language as a whole. It was frequency analysis that let Sherlock Holmes decipher this secret message, one of several at the center of Sir Arthur Conan Doyle's story "The Adventure of the Dancing Men." (The font used here differs somewhat from the original, but the message is the same):

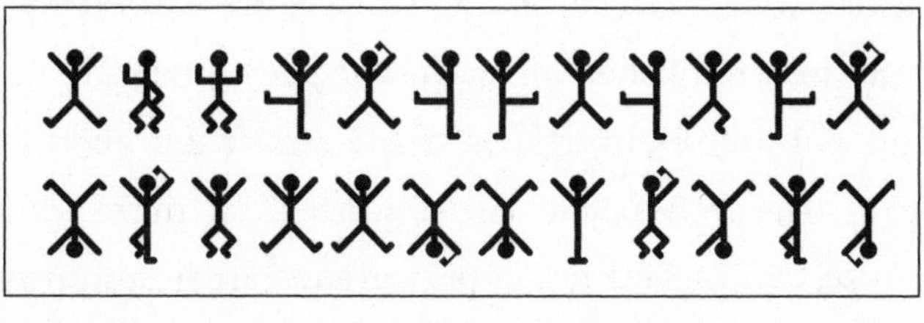

As Dr. Watson and a provincial police inspector look on, Holmes elucidates his method. "As you are aware," he says, "E is the most common letter in the English alphabet, and it predominates to so marked an extent that even in a short sentence one would expect to find it most often." To Holmes, it was immediately apparent that the character , the most frequent in the cipher, stood for "e." As for the rest of the alphabet, he continued, "Speaking roughly, T, A, O, I, N, S, H, R, D, and L are the numerical order in which letters occur." (The message, quickly unraveled, read: "Elsie. Prepare to meet thy God." Happily, Elsie did not.)

Every language has its own characteristic letter frequencies, and for decipherment, this fact can be telling. In German, the eleven most frequent letters, in descending order, are *e n i r s t a h d u l.* In French, they are *e a s t i r n u l o d.* Thus, for a simple substitution cipher like the one above, a moderately skilled investigator can use frequency analysis first to identify the underlying language (assuming it is one whose letter frequencies are known) and then to crack the cipher itself.

For frequency analysis to work properly, the text of the cipher must be long enough to provide a statistically significant sample. And that, for Kober and other investigators of Linear B, was precisely the problem: Evans, resolute as ever in old age, had continued to sit on his data. In the early 1930s, when Kober first turned her attention to the Cretan scripts, the only inscriptions to which anyone had access were the tiny handful Evans had published in *Scripta Minoa* in 1909, plus the small set published covertly by the Finnish scholar Johannes Sundwall—the "thesaurus absconditus," scholars called it. The two sets together totaled fewer than one hundred inscriptions, less than one-twentieth of what Evans had unearthed at Knossos.

With so little text available, how can a decipherer even begin? Kober began by looking for ghosts.

EVERY LANGUAGE GLIMMERS with sparks of earlier ones. These sparks—a word, a place-name—are the residual traces of languages spoken before, often long before, in the same part of the world. Though tiny, the sparks can illuminate a history of invasion, conquest, trade, and the wholesale movement of populations. In the West Germanic language known as English, we can discern Julius Caesar's invasion of Britain in the first century B.C. from linguistic survivals like *wine* (from Latin *vinum*) and *anchor* (Latin *ancora*) that remain in use today. We see the enduring presence of the Celts, who inhabited Britain before the Romans and for some time afterward, in place-names like Cornwall, Devon, and London. We also see the legacy of the Viking conquests of Britain toward the end of the first millennium: Many English words starting with *sk-*, like *skill, skin,* and *skirt,* are of Scandinavian origin. And so on.

On the same principle, as Kober knew, it should be possible to take a linguistic X-ray of a *later* Aegean language and discern traces of the lost language of the Minoans glinting beneath the surface. And so she began by scrutinizing a language she knew well, Classical Greek. Starting in the early 1930s, she spent several years combing Greek for survivals from a time before Greek speakers arrived in the region—words of pre-Hellenic origin.

Compiling an accurate list of these linguistic ghosts, Kober wrote in 1942, "would be of great importance to scholars who are trying to formulate the principles underlying the language or languages of pre-Hellenic Greece and of the Minoan scripts." These survivals by themselves would not tell her what the language of the Minoans had been—she was far too sophisticated to think that. But they might begin to reveal the structure of the words of that language. In Homer alone, she wrote, the number of pre-Hellenic words ran "into the thousands."

Among the pre-Hellenic words Kober identified in Greek were many ending in the suffix *-inth* (or *-inthos),* including *merinthos,* "cord, string"; *plinthos,* "brick" (think of the English word *plinth)*; *minthos* ("mint"); and *labyrinthos,* the "labyrinth" itself.

Her prospects improved in 1935, when Evans published his last major work, volume 4 of *The Palace of Minos.* Part of the volume was about Linear B, and it included photos and drawings of previously unseen tablets. This brought the number of available inscriptions to about two hundred. To anyone who hoped to decipher the script, that was still far from ideal, but it was a start.

For Kober, the volume's publication seemed to mark a point of no return. She had tried several times to tear herself away from Linear B, and each time found she could not. "I've resigned myself," she wrote in 1942. "If I want peace, I must first finish the job, or work till someone else finishes it." Now, at last, she could begin in earnest to compile the statistics so vital to any decipherment.

ON JULY 11, 1941, three days after his ninetieth birthday, Arthur Evans died in England, leaving nearly two thousand inscriptions from Knossos unpublished. There was little chance they would be made public anytime soon. With the outbreak of World War II in 1939, he had arranged for the tablets to be hidden for safekeeping in the Heraklion Museum. In May 1941, Crete fell to the Nazis, who quickly appropriated Evans's Villa Ariadne as their command post. At war's end, the Greek government had no money to retrieve, clean, and catalogue the tablets; they would remain in the museum, inaccessible, for years to come.

In her first public paper on the script—presented in December 1941, five months after Evans died—Kober did little to hide her resentment. "No archaeologist, however able, can be certain of finding inscriptions, but it is clearly his duty to publish them adequately when they have been found," she said. "This has not been done." As a result, she explained, scholars couldn't agree on issues as basic as how many signs Linear B contained, or precisely which signs those were. This in turn made one of the first objectives of any archaeological decipherment—compiling lists of the script's signs and words—utterly impossible. "In the sign lists published by Evans, for instance, several signs . . . appear that do not occur in any known inscription, often without the slightest clue as to their use or origin," Kober said. "Most unforgivable of all is the fact that signs actually appearing in published inscriptions do not occur in any list whatsoever." She added, in an astringent understatement: "These inadequacies hinder us at every turn."

At the time, only a few facts about Linear B could be stated with certainty, most established early on by Evans. The script direction (left to right) was known. Word divisions were readily apparent, and the numerical system was well understood. It was clear that Linear B was a syllabary. The script also contained a set of logographic characters, and the meaning of many of these was plain. That the Cretan scribes used separate pictograms for male and female animals was known, though scholars disagreed as to which represented which. This was also the case for the words "boy" and "girl"—𐀒𐀺 and 𐀒𐀷—identified as a pair in 1927 by the historical linguist A. E. Cowley.

The meaning of only one Linear B word was known with any degree of certainty. The word was 𐀓𐀫, and while no one knew how to pronounce it, context revealed its identity: It appeared regularly at the bottom of Minoan inventories, just before the tally, and almost assuredly meant "total." All in all, it was not much to show for forty years' work.

HOW DOES A decipherer penetrate such a tightly closed system? There is only one way possible, and it can be illustrated by means of a delightful puzzle from the International Linguistics Olympiad, an annual competition for high school students.

The puzzle, adapted below from one used in the 2010 contest, was created by Alexander Piperski; it involves a real writing system, known as Blissymbolics. Blissymbolics was invented after World War II by Charles K. Bliss (né Karl Blitz), an Austrian Jew who had survived Buchenwald. It is an attempt to devise a universal writing system that can be understood by speakers of any language. Blissymbolics is an example of a pure ideographic system, in which each symbol stands for an entire concept.

Here are eleven English words, numbered for reference, written in Blissymbolics:

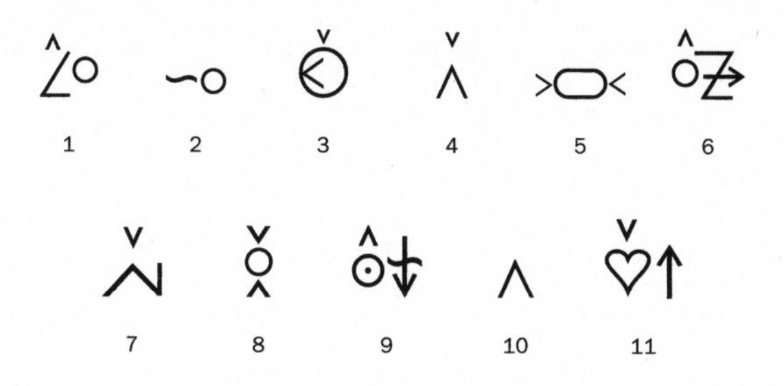

The translations—in random order—are these:

waist	active	sick	lips
activity	to blow	western	merry
to cry	saliva	to breathe	

The object is to match each symbol with the correct translation. (The general method of decipherment is described below; the full solution is revealed in the Notes on page 317.)

When one confronts a problem like this, a natural first reaction is panic: The strange symbols look utterly impenetrable. But they aren't, as closer inspection reveals, for every clue needed to solve the puzzle lies buried somewhere within the problem itself.

From the outset, we have one great advantage over the Linear B investigators: We are dealing here with a bilingual inscription—albeit one with the translations in the wrong order. Even so, this gives us a tremendous head start. Let us begin, then, by examining the set of English words.

We see immediately that the set contains different parts of speech: nouns, adjectives, and verbs. This is the first internal clue the problem offers, and we will make use of it in the time-honored way, by counting. We arrive at this tally:

4 nouns: waist, lips, activity, saliva
4 adjectives: active, sick, western, merry
3 verbs: to blow, to cry, to breathe

These numbers may prove useful: We now know that the group of eleven Bliss symbols contains four nouns and four adjectives, but only *three* verbs. If we can find something that characterizes precisely three of the symbols, we may well have found the verbs. With this in mind, we turn now to the symbols themselves. At this stage, we won't try to translate any of them, however tempting that might be. We are going to treat them as objects of pure form and nothing more.

As we study the symbols, we notice two things. First, all but one of them (No. 10) have more than one component part. This may be important, though we don't yet know how. Second, several symbols are topped with smaller marks—a hatlike caret (ˆ) or a V-shaped sign (ˇ). We can sort the eleven symbols on this basis:

3 topped with caret: Nos. 1, 6, and 9
4 topped with V-shape: Nos. 3, 4, 7, and 11
4 plain symbols: Nos. 2, 5, 8, and 10

We have now made our first major discovery. We already knew we were looking for three verbs: In the three symbols topped with carets, we may have found them. Let us tentatively designate these symbols as verbs. We still don't know which symbol corresponds to which verb, but for the time being that doesn't matter. The ground we have gained so far can be summarized thus:

∠̂o, ôZ⃫, and ⊙̂↓ = possible verbs (order unknown): "to blow," "to cry," "to breathe."

Now we seem to get stuck. We are left with four symbols topped with the V-shaped mark, and four plain symbols. There is no way to tell which group contains the nouns and which contains the adjectives. Or is there? Once again, we must scour the symbols for internal clues.

If we turn back to the English translations, we notice something tantalizing. Two words in the list are clearly related to each other: the noun "activity" and the corresponding adjective "active." They differ in *only a single parameter*: part of speech. It would help enormously if we could find two Bliss symbols that are alike except for one parameter.

Scanning the symbols, we find just such a pair: 4 and 10, Λ̌ and Λ. One has the V-sign; the other doesn't. One is the adjective "active," the other the noun "activity." The V-sign either turns a noun into the corresponding adjective, or an adjective into the corresponding noun. But which? To answer the question, we must now start to make educated guesses about the meanings of individual signs, much as Champollion guessed that the hieroglyphic ⊙ was meant to invoke the Coptic word "sun."

Let's begin by examining the four characters with V-shaped tops, Nos. 3, 4, 7, and 11: ⓒ̌, Λ̌, Λ̌ı, and ♡̌↑. If they are nouns, then they mean (in random order) "waist," "lips," "activity," and "saliva." If they are adjectives, then they mean (also in random order) "active," "sick," "western," and "merry."

To the layman's eye, one sign in this group leaps out: No. 11, the heart with a V-shaped top and an upward-pointing arrow. We know that Blissymbolics is an ideographic system, meant to be understood universally. In many cultures, the heart is considered the seat of emotion. Does any of the remaining English words suggest emotion?

One does: the adjective "merry." If we assume that symbol No. 11 means "merry," we can then forge the following deductive chain:

- If ♡̌↑ = "merry," then the V-sign turns nouns into adjectives.
- Therefore, Λ = "activity," and Λ̌ = "active."
- The three remaining nouns, ⌐o, >⊂⊃<, and ǒ/λ, are (order unknown) "waist," "lips," and "saliva."
- The two remaining adjectives, ⊙̌ and Λ̌, are (order unknown) "sick" and "western."
- The verbs ∠̂o, ôZ, and ô↓, are (order unknown) "to blow," "to cry," "to breathe."

The rest of the symbols can be unraveled in similar fashion, until the meaning of all eleven is established. It will help to note that many of the English words refer to parts of the body or bodily functions. It will also help to think about the multipart form that most of the symbols take: This form may well encode a meaning that is the sum of the individual components.

CONFRONTING THE LINEAR B inscriptions, Kober and the other analysts faced a similar deductive process, gridded up a thousandfold. But to her great disgust, most investigators persisted in looking at the problem through the wrong end of the telescope, seeking first to identify the language the Minoans spoke and only afterward to unravel the script.

Everyone, or so it seemed, had a theory about what language the tablets recorded. Michael Ventris, for instance, who had begun thinking fervently about the problem as a schoolboy, was convinced it was the lost Etruscan tongue. He clung steadfastly to the idea until weeks before his decipherment, when he was forced to contemplate a vastly different scenario. Others held even stranger notions. "It is possible to prove, quite logically, that the Cretans spoke any language whatever known to have existed at that time—provided only that one disregards the fact that half a dozen other possibilities are equally logical and equally likely," Kober said at Yale in 1948. "One of my correspondents maintains that they were Celts, on their way to Ireland and England, and another insists that they are related to the Polynesians of the Pacific."

Kober indulged in no such speculation, remaining firmly agnostic on the language of the script. ("I am interested," she once said, "only in what the Minoans wrote.") The tablets, she insisted, must be analyzed based on internal evidence alone, and the scripts must be allowed to speak for themselves, absent the decipherer's prejudice.

"We cannot speak of *language,* but only of *script,* because the system of writing cannot be read," she once said. "Since all we know about the language or languages of the scripts is what the graphic remains show, a thorough understanding of these written documents is necessary before any linguistic theories are promulgated. Otherwise it is impossible to avoid reasoning in a circle."

By the mid-1930s, with two hundred inscriptions at her disposal, Kober could begin to sift the teeming mass of symbols for the kind of internal clues she sought. And so, hour by hour and symbol by symbol, she began to count.

* * *

A LANGUAGE AND a writing system are not remotely the same thing, though each impinges on the other revealingly. In any given language, sentences can be deconstructed into words, and words further deconstructed into smaller component parts. These parts occur, importantly, "in patterns of selection and arrangement," as the anthropologist E. J. W. Barber has said. The script used to write that language will display corresponding patterns.

A simple example: In any inscription, Barber writes, "each sign bears a relation to the signs adjacent to it." In Italian, for instance, it is permissible for *s* to be followed by *f* or *g* at the start of a word: *sforza* ("force"), *sfumato* ("smoky"), *sgraffito* ("drawing, writing"; compare English *graffito* and its plural, *graffiti*). Not so in English, for no reason other than that is simply how English chose long ago to behave. As a result, certain characters will crop up side by side in certain positions in written Italian but not in written English. Tabulating such behavior helps the decipherer seeking the language of an inscription narrow the field of likely suspects.

When Kober began her work, she kept her tabulations in a series of notebooks. In just a few years, she would fill forty of them—the twentieth-century account books in which she itemized the Bronze Age account books of the Minoan kingdom. But during World War II and afterward, paper was scarce, and she had to resort to enterprise (and occasional genteel larceny) to keep her statistics going. When she could no longer get notebooks, she began hand-cutting two-by-three-inch "index cards" from any spare paper she could find: church circulars, the backs of greeting cards, examination-book covers, checkout slips from the college library, and whatever else she could lay her hands on.

On the front of each card, Kober inked statistics for various

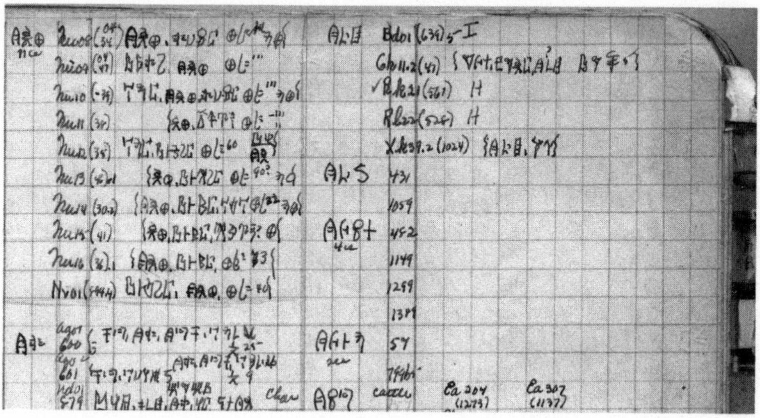

Detail of a page from one of Alice Kober's sign-juxtaposition notebooks. At upper left, she is analyzing instances in which the sign-group [Linear B signs] is followed by other sign-groups with the same first character. Note the handmade tabs with Linear B characters down the right-hand side of the page.

signs in the minutest of writing. On the backs of the cards, bits of residual text, like fragmentary inscriptions on broken tablets, gave evidence of their original use:

> Four Lenten Lectures
> Thursday at 8:15 P.M.
>
> The Barbizon-Pla

> o pages are to be torn from this book.
> ll matter not intended for correction
> y the teacher should be crossed out
> y the student, but the book should

Before her death, Kober would cut and annotate more than 116,600 two-by-three-inch slips, as well as more than 63,300 larger slips—some 180,000 cards in all. The smaller ones she fitted neatly into empty cigarette cartons, the one paper product of which she appeared to have no short supply. Even now, more than six decades later, to open one of them is to be met with a faint whiff of midcentury tobacco. FLEETWOOD IMPERIALS: A CLEANER, FINER SMOKE, her ersatz file cabinets say. HERBERT TAREYTON: THERE'S SOMETHING ABOUT THEM YOU'LL LIKE.

A WRITING SYSTEM is a woven fabric, an interlaced network of sounds and symbols. As with real woven cloth, there are many different interlacements possible—many different ways, that is, in which sound and symbol may be bound up together. When a decipherer confronts an unknown language in an unknown script, she must hunt for a thread that, pulled, will begin to reveal the structure of the weave. She does this by searching for patterns, for every combination of language and script, as Barber writes in *Archeological Decipherment,* bears "a distinctive fingerprint . . . which may help us recognize the language involved." It was for this reason that Kober had done fifteen years' preparation, learning languages that employ diverse array of scripts, from logographic (Chinese) to syllabic (Akkadian) to alphabetic (Persian).

Her task would have been so much easier if only the Minoans had used an alphabet. A syllabary is far harder to decipher, because in a syllabary nearly every character stands for more than one sound. As a result, frequency counts skew very differently for syllabic scripts than they do for alphabets, and the same language will behave very differently statistically written with the one than with the other. A two-syllable word written syllabically may entail precisely two signs. Written alphabetically, as Barber notes, the same word "may be represented by from two to a dozen signs"—think of the vast difference in length between two-syllable English words like *Io* and *strengthened.*

As Kober was painfully aware, there were no published frequency tables for syllabically written languages. That is where her homemade file cards came in, helping her corral the seemingly random mass of Linear B characters into orderly statistical sets. For every word on the tablets—the two hundred published inscriptions gave decipherers about seven hundred different words to work with—she cut a separate card. On each card, in her tiny hand, she recorded as much data as she could mine about the characters it contained.

She catalogued the frequency of each character, of course, but she catalogued a great deal more than that. She noted the frequency of each character in any *position* in a word (initial, second, middle, next-to-last, and final); the characters that appeared before and after every sign; the chances of a given character's occurring in combination with any other character; repeated instances of two- and three-character clusters; and much else.

In a letter to a colleague in 1947, Kober itemized the time it took her to compile a single statistic, the one that tallied a sign's frequency in combination with others: "You can figure out for yourself how long it will take to compare each of 78 signs with 78 other signs, at 15 minutes (with luck) for each comparison. Let's see, 78 times 77 times 15 minutes—that's about 1,500 hours. I did it on the little slide rule I just bought to hasten the arithmetic I'll have to do."

After calculating her figures, Kober, using a dime-store hand punch, made holes in precisely determined spots on her cards. Each spot stood for a particular Linear B character: There was a designated spot for [Linear B character], for instance, another for [Linear B character], another for [Linear B character], and so on. The location of each character's hole remained constant from card to card. In this way, by stacking two or more cards together, Kober could see which holes aligned. That told her instantly which characters those words had in common. "Making all these files takes time, and the files have considerable bulk," she wrote in 1947, "but once they are finished, all the material is so arranged that anything, no matter how unexpected, can be checked in a few minutes."

What she had created, in pure analog form, was a database, with the punched holes marking the parameters on which the data could be sorted. But for all their rigor and precision, the file boxes also "reveal a gentler side to Alice Kober," as Thomas Palaima and his colleague Susan Trombley have written. On one occasion, they note, Kober "took extra care in cutting a greeting card used as a tabbed divider, perfectly centering a fawn lying in a bed of flowers."

ALL THE WHILE, Kober was shouldering a full course load at Brooklyn College. "I . . . teach in my spare time, so to speak," she wrote with bitter humor to a colleague in 1942, and she was scarcely joking. The college was part of New York City's public university system, and for faculty members of that period, the emphasis was on teaching rather than research. "Brooklyn

College never did anything for me in a scholarly way," Kober lamented a few years later. She shared an office with four other people (this alone would have made on-campus scholarly work impossible); in addition to her classes she was active in the customary round of university work, serving on a spate of faculty committees.

Two of Alice Kober's cigarette-carton card files, with the ersatz "index cards" on which she sorted the characters of Linear B.

In 1943, asked to give private instruction in Horace to a blind student, Kober taught herself Braille—itself a writing system—and from 1944 onward, she brailled textbooks, library materials, and final exams for all the blind students at the college. It took as many as fifteen hours to braille a single exam.

In a letter to her department chairman in May 1946, Kober itemized her schedule for the coming weeks:

May 21–24.	brailling examinations; preparation for a short speech to be given May 24. . . . Classes.
May 25–June 6.	Classes; braille.
June 7.	Examination in Latin 01; proctoring.
June 7–10.	Marking papers; marks due on Monday, June 10.
June 11.	Additional proctoring assignment in the department.
June 12.	Examination in Classical Civilization 1 (110 papers). Proctoring.
June 13.	Marking papers.
June 14.	Examination in Classical Civilization 10; proctoring. (30 papers).
June 15.	Morning—marking papers for the New York Classical Club scholarship. Evening—Phi Beta Kappa initiation.

"The less said about teaching assignments, the better," Kober complained to a fellow classicist in 1949. "Blankety-blank-blank will have to do. Too bad we can't get paid for doing what we want to do."

But with her pupils, she was warm and involved, even quietly passionate. In the early 1940s, at her students' request, she established the Hunter chapter of Eta Sigma Phi, the national classics honor society, and was its faculty adviser. When she taught archaeology, she took her students excavating in nearby vacant lots, where they unearthed cutlery and broken dishes.

"I think I'm a good teacher, at least my students come to class with a smile, laugh at my jokes," Kober wrote in 1947 when applying for a prestigious job at the University of Pennsylvania. Her appraisal was seconded by at least one grateful student. Among Kober's papers is a letter she received in 1944 from a young woman in Detroit named Fritzi Popper Green:

> *Dear Miss Kober,*
>
> *I know that* [*the*] *name and address at the top of the page mean nothing to you. . . . However, my identity is not really important. I was just one of that lucky group of Latin students who, during less troublesome times some six or seven years ago, enjoyed Horace and Plautus and Terence under your capable guidance in the evening session at Brooklyn College.*
>
> *I wanted you to know how much it meant to me when you carried us through so that we had sufficient credit to consider Latin our major—despite the fact that you no longer wanted to teach at night. . . .*
>
> *I do want you to know, even at this late date, that I was one student who never believed or considered that Latin was "not practical" and that whatever love and understanding I have for the classics, I attribute, for the most part, to you.*

It was clear that in her "soberly undramatic" way, Kober transmitted to her students the passion she felt for the life of the mind. Her former pupil Eva Brann remembered her telling the class, "You know a great work when the back of your neck tingles."

IN LATE 1945, with Linear B beckoning, Kober applied to the Guggenheim Foundation for one of its coveted fellowships. If she was good enough to get it, she would have a year free from teaching for the first time in more than a decade—a blissful year alone with the script. Perhaps then she could make real headway toward the day when the ancient Cretan tablets could be read once more. It was not the content of the tablets that interested her per se, but the riddle of the script as a pure cryptographic problem. As she told the young Phi Beta Kappa initiates at Hunter that June evening, the decipherment of the Minoan tablets might not yield high drama, but to the decipherer, it would bring satisfaction of the deepest kind anyway:

> We may find out if Helen of Troy really existed, if King Minos was a man or a woman, and if the Cretans really had a mechanical man who marched along the cliffs of Crete and warned the inhabitants when hostile sea-farers tried to land. *On the other hand, we may only find out that Mr. X delivered a hundred cattle to Mr. Y on the tenth of June, 1400 B.C.* But that is one of the hazards involved. After all, solving a jigsaw puzzle is no fun, if you know what the picture is in advance.

5

A DELIGHTFUL PROBLEM

IN APRIL 1946, THE NEWSPAPERS announced the 132 recipients of that year's Guggenheim Fellowships. There were luminaries on the list, including the photographer Ansel Adams, the poet Gwendolyn Brooks, and the composer Gian Carlo Menotti. Also on the list, receiving the first major award of her career, was Alice Kober. The $2,500 stipend would free her from teaching for twelve months, time she could devote without interruption to Linear B.

Congratulations streamed into Brooklyn. Sir D'Arcy Wentworth Thompson, a Scotsman who was, in Kober's words, "one of the 'grand old men' of classical scholarship," wrote her, "Your learning is great, your courage is immense, and your problem is altogether delightful." Leonard Bloomfield, the most eminent linguist of the first half of the twentieth century, sent Kober an even more arresting tribute. He wrote: "The very notion of your problem scares me."

Kober arranged for a year's leave from Brooklyn College, to begin on September 1. As the foundation required, she underwent a physical exam; in light of the devastating illness that would overtake her in little more than three years' time, it is heartbreaking to read her doctor's report today. "I find she is in excellent physical condition," he wrote. "There is no evidence of any organic [or] functional disease."

Even before her fellowship began, she had much to keep her busy. She started work on a major review, for the *American Journal of Archaeology,* of volume 1 (there would eventually be two) of Bedřich Hrozný's "decipherment" of Linear B. Hrozný, the decipherer of Hittite cuneiform, was determined to show that the language of Bronze Age Crete was also a form of Hittite. "It seems the sheerest balderdash," Kober wrote privately to John Franklin Daniel, the journal's new editor. "But Hrozný has been lucky with Hittite . . . so I want to consider very carefully before I make the nasty comments I feel like making."

The more Kober considered, the more furious she grew at Hrozný's unscientific approach to the script. "I hope he will not be too annoyed with my review," she wrote to another colleague, "but I feel that in scholarly matters the truth must always be told." By the time she sat down to write the review, she was so inflamed that she told Daniel he would need asbestos paper on which to print it. "Don't cool off too much before writing your review of Hrozný," Daniel replied cheerfully by return mail. "I have checked and find that we have a small stock of asbestos paper available."

Daniel, a young archaeologist at the University of Pennsylvania, was one of the few people in Kober's field whom she genuinely respected. (Ventris, by contrast, was not.) In 1941, Daniel had published an important article on the Cypro-Minoan script, a Late Bronze Age writing system used on Cyprus and thought to have been derived from Linear A. He and

Kober began corresponding in the early 1940s, after Daniel succeeded Mary Swindler as the *AJA*'s editor; by the middle of the decade, he would become the most important person in Kober's professional life.

The months before her fellowship also let Kober expand her linguistic repertory in preparation for Linear B. She had long since mastered a host of languages ancient and modern: Greek and Latin, of course, as well as French and German, all standard issue for a working classicist, plus Anglo-Saxon. Starting in the early 1940s, she had set about learning Hittite, Old Irish, Akkadian, Tocharian, Sumerian, Old Persian, Basque, and Chinese. From 1942 to 1945, while teaching full-time at Brooklyn, she commuted weekly by train to Yale to take classes in advanced Sanskrit. She knew far better than to expect to find any of these languages lurking in the tablets. What she was doing, as she made clear in her correspondence, was arming herself with them "against the happy day when they may do me some good."

Kober had also prepared by studying Old and New World field archaeology: In the summer of 1936 she took part in an excavation in Chaco Canyon, New Mexico; in the summer of 1939, she explored ancient sites in Greece. Years later, she would write of being "homesick for Athens," wistfully recalling "my scrambles up and down the north slope of the Acropolis."

She did manage a vacation of sorts before her fellowship began, driving with friends through the American West and down into Mexico in the summer of 1946. (Her brief account of the trip is one of only two passing references to social life in her entire archived correspondence, which comprises more than a thousand pages.) But as a letter she wrote to Daniel mid-voyage, on hotel stationery from Boulder, Colorado, indicates, she took Linear B along.

It was so like Kober to do two things at once, squeezing as much as she could from every available moment. For relaxation, she occasionally knit, or read detective stories. Unlike most people, however, she did them simultaneously. The mathematical elegance of each pursuit must have appealed to her greatly, as did the possibility of their efficient combination.

ON SEPTEMBER 1, 1946, Kober's Guggenheim year officially began. She spent the first several months "clearing up the background," as she called it—analyzing more Asia Minor languages like Lydian, Lycian, Hurrian, Hattic, and Carian "and some remnants of others." Many of these had no published word lists available, and she had to make card files for each language from scratch. "It's a thankless job, and most scholars wouldn't undertake it," she wrote Daniel that fall. "As a result, every person working in the field must do all the work over again." For Lydian alone, she said, it took "about a month of extremely intensive work" just to set up her files.

Nonetheless, her joy in the freedom from teaching was palpable. "I had never before been able to work uninterruptedly for such long periods," Kober wrote in a progress report for the Guggenheim Foundation in December. "It takes about a month to do what was formerly a year's work." In closing, she added, "I only hope my results equal my gratitude."

To a great extent, those results would depend on the reply to a letter she had written the month before, in November 1946. A model of diplomacy, yet shot through with barely concealed yearning, it was something she had been steeling herself to compose for some time.

The letter was to Sir John Linton Myres, a distinguished archaeologist at Oxford University. Forty years earlier, Myres had been Evans's young assistant at his excavations of Knossos. After Evans's death in 1941, Myres inherited from him the mantle of grand old man of Aegean prehistory. Unfortunately for all concerned, he also inherited Evans's work on Linear B, and it was he who was charged with putting four decades of notes and transcriptions into publishable shape. Evans, as Andrew Robinson observes in *The Story of Writing*, had left "a disorganized legacy." His planned second volume of *Scripta Minoa*, devoted to Linear B, never materialized in his lifetime. The thankless task of producing it now fell to Myres, who by the mid-1940s was elderly and ill himself.

When Kober wrote to him, scholars still had access to only two hundred inscriptions. Myres had Evans's transcriptions and photos of nearly two thousand more: Evans had decreed that they were to remain unseen until *Scripta Minoa II* was published. If Kober could see them privately beforehand, her store of inscriptions would increase almost tenfold. But knowing Evans's desire for control of his material—even, it would seem, from beyond the grave—she held out little hope.

"Dear Professor Myres," she wrote on November 20, 1946:

> *This letter is an extremely difficult one to write because I have a request to make which I fear cannot be granted. Yet, under the circumstances, it is necessary for me to ask. . . .*
>
> *If it is in any way possible for me to see some, or all, of the unpublished inscriptions during the year when I am free to*

> *devote all my time to them, it would be of great advantage to me, and perhaps would speed the ultimate decipherment. . . .*
>
> *I had hesitated to impose the request upon you, because I know publication of the Corpus is under way, and although it is now over fifteen years since I began working on the problem, I have always felt that Sir Arthur Evans should be the one to decipher the scripts, since he was the one to whom we owe the knowledge of their existence. If it was his wish that no one should see the unpublished inscriptions until they were formally published under his auspices, I cannot help but agree that he deserves that tribute to his memory.*
>
> *Since, however, scholarly interest sometimes resembles religious fanaticism in its demands, I feel I must ask you whether I could see the unpublished material during my year of freedom from academic duties. . . .*

In closing, she wrote, with characteristic quiet ardor, "I must confess, strong as the statement sounds, that I would gladly go to the ends of the earth, if there were a chance of seeing a new Linear Class B inscription when I got there."

She signed the letter, "Apologetically, yet hopefully, (Miss) Alice E. Kober." Then she settled down with her cigarettes and her slide rule to wait for an answer.

BY THEN, KOBER—working with only two hundred inscriptions—had already taken a significant step forward. The thousands of hours spent sifting her cards had begun to pay dividends, and, piece by piece, the picture on the puzzle box was starting to emerge. In 1952, in his triumphant announcement of the decipherment on BBC Radio, Michael Ventris discussed the steps by which a decipherer makes his way through the forest of symbols:

> The usual way of putting the signs of a syllabary into some standard order, when we know how they are pronounced, is to arrange them on a syllabic grid. . . . The most important job, in trying to decipher a syllabary from scratch, is to try to arrange the signs provisionally on a grid of this sort, even before we can work out the actual pronunciation of the different vowels and consonants. . . . Once we can determine, later on, how only one or two signs were actually pronounced, we can immediately tell a good deal about many other signs which lie on the same columns of the grid . . . and it can only be a matter of time before we hit on the formula which solves it.

What Ventris did not say was that it was Alice Kober, at her dining room table, who had first done those very things.

In 1945, Kober published what would be the first of her three major articles on Linear B. Her great contribution to the decipherment—elegantly on display in all three papers—lay not only in *what* she found but also in the means *by which* she found it. Several things about her method stood out:

To begin with, she assumed almost nothing, disregarding nearly every idea about the script that had been put forward by others. (Practically the only thing Kober did not doubt was that Linear B was a syllabary. "If all the Minoan scripts aren't syllabic, I'll eat them," she wrote privately. "That's the one thing I'm sure of.") Second, she refused to speculate on the language of the tablets. Third—and this was the most radical departure of all—she refused to assign a single sound-value to the Linear B characters.

One of the greatest temptations an archaeological decipherer faces is to assign a phonetic value to every character at the start—to speculate, that is, on which sounds of the language the various characters stand for. With an unknown language in an unknown script, there is little way to do this besides random guesswork. The great danger of doing so, as Kober repeatedly warned, is that it sets up a cycle of circular reasoning: The characters of a script correspond to the sounds of Language X; therefore, the script writes Language X.

To Kober, assigning sound-values at the outset was the refuge of the careless, the amateurish, and the downright deluded. By contrast, she treated the symbols of Linear B as objects of pure form, looking for patterns that might lead her, all by themselves, into the structure of the Minoan tongue. In the Blissymbolics problem above, we began with form *and* meaning: the curious symbols and the set of English translations. In Linear B, investigators faced a far more brutal situation: They were forced to inhabit, as Kober evocatively wrote, a world of "form without meaning." Of all the would-be decipherers, she was the one most willing to dwell there for as long as it took.

Methods like these inform all of Kober's writing about Linear B. They come to full flower in her three important papers on the script, published in the *American Journal of Archaeology* between 1945 and 1948. In her 1945 paper, "Evidence of Inflection in the 'Chariot' Tablets from Knossos," she examined a series of tablets, reproduced by Evans in *The Palace of Minos,* that deal with chariots and their constituent parts. (The "Chariot" tablets are noteworthy among the Knossos tablets for containing complete sentences of the Minoan language rather than mere word lists.) The tablets in question, as hand-copied by Kober from Evans's book, included these:

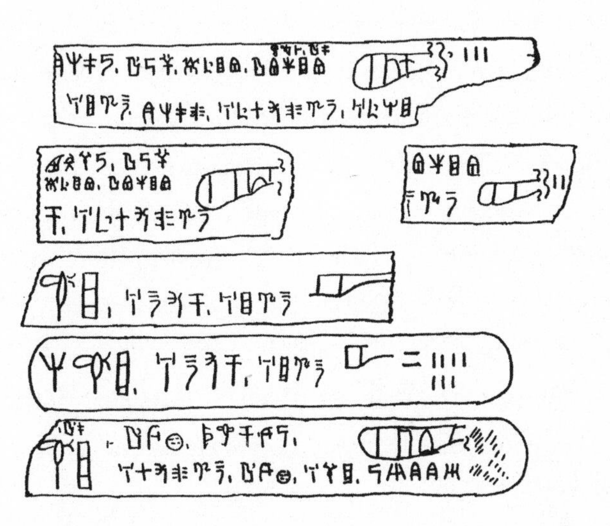

Kober's paper was significant for two things, both of which would have a profound effect on the course of the decipherment. First, it showed that it was possible to analyze the Cretan symbols without attaching sound-values to them. Second, it proved that the language of Linear B was *inflected*—that is, that it relied on word endings, much as Latin or German or Spanish does, to give its sentences grammar. Though on its face this discovery may look like a small find, it would eventually furnish the long-sought "way in" to the tightly closed system of Linear B.

Roughly speaking, inflections are word endings, also

known as suffixes. Languages make use of inflection to varying degrees: In English, inflections include *-s* (which indicates a third-person singular verb); *-ed* (past tense); *-ing* (participle); *-s* (again, this time indicating a plural noun); and a few others. Modern English has only about eight inflections. Other languages, like Chinese, have almost none.

Some languages, however, use far more, as anyone who has sweated over Latin, with its constellations of word endings, can attest. In Latin, verbs can be inflected for a spate of grammatical attributes, including person (first, second, or third person) and number (singular or plural), depending on who is performing the action of the verb. Tiny but powerful, these inflections impart the kind of information that in English usually requires extra words. Consider the Latin verb *laudāre,* "to praise." When any of the following suffixes is attached to its stem *(laud-),* the result is an entire small sentence, elegantly bound up in a single inflected word:

	Singular	Plural
1st person:	*lau**dō**,* "I praise."	*lau**dāmus**,* "We praise."
2nd person:	*lau**dās**,* "You praise."	*lau**dātis**,* "You-all praise."
3rd person:	*lau**dat**,* "He/She praises."	*lau**dant**,* "They praise."

The nouns of Latin are also richly inflected. In the table below, inflections mark the noun's "case," telling us what role it plays in a given sentence: subject, direct object, indirect object, possessor, or addressee:

Case	Singular	Plural
Nominative (subject):	***porta***, "the gate"	***portae***, "the gates"
Genitive (possessor):	***portae***, "of the gate"	***portārum***, "of the gates"
Dative (indirect object):	***portae***, "to/for the gate"	***portīs***, "to/for the gates"
Accusative (direct object):	***portam***, "the gate"	***portās***, "the gates"
Ablative (movement away from):	***portā***, "away from the gate"	***portīs***, "away from the gates"
Vocative (addressee):	***porta***, "O gate!"	***portae***, "O gates!"

For the analyst of an unknown script, demonstrating that the language of the script is inflected is a major diagnostic triumph, for it narrows the field of likely candidates in a single stroke: One can immediately eliminate inflection-poor languages like Chinese and focus the search on inflection-rich ones like Latin. But there is a catch: Conclusive proof of inflection is extremely hard to come by—much harder than it seems.

When Evans first studied the Knossos tablets, he noticed small sequences of symbols that seemed to recur at the ends of words. Among them were these:

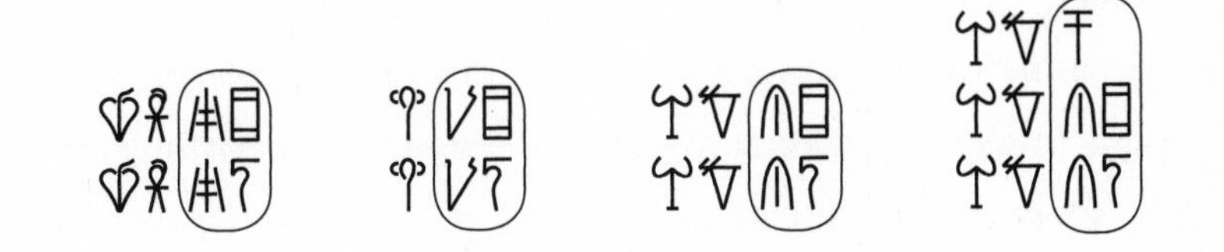

He decided that these sequences were inflections on the ends of Minoan nouns. (He thought they indicated gender, or perhaps case, much as suffixes on German and Latin nouns do.) But inflections are wily things, and it is all too easy for an investigator to spot them where they don't actually exist. Consider the following four English words, written in the Dancing Men font to make them more cipherlike:

A decipherer will immediately spot the three-character sequence at the end of each word. It certainly looks like a suffix, perhaps *-ing*. "Eureka! Inflection!" he cries. But is it really?

Deciphered, the four words turn out to be these:

= finger
= wager
= lager
= tiger

Here, "inflection" is a false friend, for as any English speaker knows, there is no suffix *-ger* in the language; nor are there stems *fin-* (in this sense), *wa-*, *la-*, or *ti-*. The repeated pattern at the ends of these words is mere coincidence and nothing more.

Had Evans truly found inflection in Linear B, or had he been similarly seduced? Though he wrote of his little paradigms, "We have here, surely, good evidences of declension" (that is, inflection on the ends of nouns), his discovery was intuitive and circumstantial, and had to be taken on faith. Kober knew that for the decipherment to advance, someone had to prove conclusively whether or not the Minoan language was inflected. This she set out to do.

If you are studying an unknown language in an unknown script and you see what looks like inflection, how do you prove it actually *is* inflection? For Kober, the answer was to be found in the "patterns of selection and arrangement" that had emerged over the years as she pored over her cards. She had let the tablets speak for themselves and now they were beginning to reward her.

"If a language has inflection, certain signs are bound to appear over and over again in certain positions of the written words, as prefixes [or] suffixes," she wrote in her 1945 article. "No matter how much these changes may be obscured, the fact that they occur regularly must reveal them, if the amount of material available for analysis is large enough, and the analysis sufficiently intensive."

As she studied the "Chariot" inscriptions, Kober noticed certain words, and even phrases, that recurred on two or more tablets. Among them was the word [Linear B signs], which appeared on two tablets; [Linear B signs], which appeared on three; and [Linear B signs], found on five. She also identified an entire three-word phrase: [Linear B signs], which was repeated on five different tablets.

Strikingly, Kober also found words *that recurred in similar but not quite identical forms*—forms that differed only in their last few characters. The words seemed to have the same stem but different suffixes. They included [signs] and [signs] (where the stem starts with [signs]); [signs] and [signs] (where it starts with [signs]); and [signs] and [signs] (where it starts with [signs]).

It was also significant, as she noted, that a variety of Minoan endings could combine with a variety of stems, a defining property of suffixhood. (In English, for instance, the suffixes *-er* and *-ing* can attach to a wide array of stems, including *sing-*, *teach-*, and *work-*. They cannot, by contrast, attach to "pseudo-stems" like *wa-*, *la-*, and *ti-* above.)

A similar process, Kober showed, could be seen on the Linear B tablets. The suffixes -[sign] and -[sign], for instance, could combine not only with the stem [signs], but also with other stems, as attested by the presence of word pairs like [signs] and [signs]; [signs] and [signs]; and [signs] and [signs]. Strikingly, although Kober had not set out looking for them, the suffixes -[sign] and -[sign], which she found so often at the ends of Linear B words, were also the ones Evans had instinctively noted:

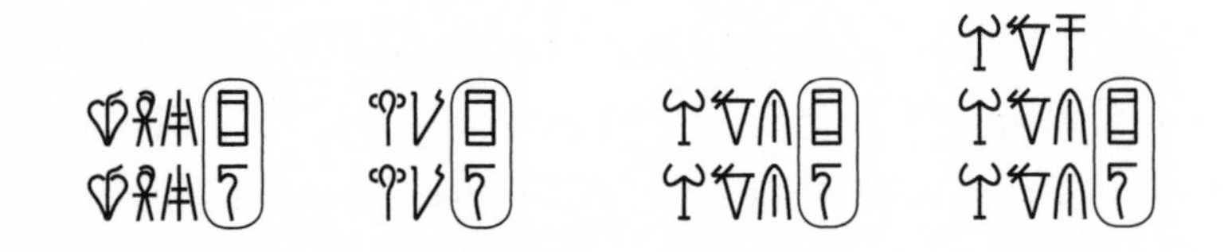

What Kober showed was not merely that these suffixes alternated in Linear B words, but also that they alternated *in regular fashion*, attaching to the same range of stems, just as *-er* and *-ing* do in English. "It was one thing to suggest that the writing on the Linear B tablets might conceal an inflected language," Maurice Pope wrote in *The Story of Archaeological Decipherment*. "It was quite another to establish definite patterns of inflection. This is what Miss Kober did."

It was a pathbreaking discovery, and it accomplished several things. For one, it narrowed the field of possible candidates for the language of the tablets. For another, it settled one vexing problem immediately: the question of whether the languages of Linear A and Linear B were the same. It was clear that the Linear B *script* had developed from the earlier Linear A; on that, scholars could agree. But were the languages they wrote also related? Most investigators, including Evans and Ventris, believed they were.

But to Kober, the data told a far different story. Her preparation had also included careful study of the published Linear A inscriptions, and none of them showed evidence of inflection. The conspicuous presence of inflection in Linear B, and its conspicuous absence in Linear A, convinced her that the respective languages were different. Hers was the minority view, but one to which she held firmly.

Finding inflection in Linear B would lead Kober to even deeper discoveries about the script, which she would publish toward the end of the 1940s. These discoveries centered on the very particular things that happen when an inflected language is written with a syllabic script. This critical interaction, she showed, can offer vital clues to the nature of the language in question. She hinted as much in 1945, in a telling footnote to her first major paper:

"When a syllabary is used [inflections] are bound to be obscured," she wrote, "since even the simplest syllabary consists of signs that combine a consonant and vowel, so that even a slight inflectional alteration may produce a word spelled with entirely different signs." It was this observation, more than anything else, that would be the linchpin of the decipherment.

* * *

ON DECEMBER 13, 1946, a letter with an Oxford return address arrived at Kober's home. She let an hour go by before she dared open it. When she did, she was astounded: Sir John Myres had granted her request to see the Knossos inscriptions. "Please let me know your plans and let me know how I can further them," he wrote. "The whole material is in my control."

"I cannot tell you how happy your letter made me," Kober wrote him in reply. "My fondest hopes [have] materialized." But as things played out, getting access to the inscriptions was both the best and the worst thing that could have happened to her.

6

SPLITTING THE BABY

"IN A LITTLE OVER A month, I'll be in the same town with the *Scripts*," Kober wrote rapturously to Myres in early February 1947. She had just booked her passage by sea—she was somewhat afraid to fly—and would sail from New York in March. On arriving in England, she would travel by train to Oxford, where she would live at St. Hugh's, one of the university's women's colleges.

In the meantime, some arduous preparation lay in store for her. Kober would have just five weeks in Oxford—five weeks in which to copy nearly two thousand inscriptions from Evans's photographs and drawings. To make matters more difficult, she would have to copy everything by hand: the ubiquitous office photocopier was years in the future. As she must have felt keenly, when it came to the efficient duplication of written text by a lone copyist, the techniques available to scholars in the mid-twentieth century had not advanced much beyond those employed by the Cretan scribes three thousand years before.

What techniques there were tended toward the shoddy. Kober's hand-cut index cards testify to the continued scarcity of paper years after the war ended; the little that could be had was usually substandard. Several times in her correspondence of the late 1940s she vents her frustration at the available paper, which was of such poor quality that it would barely take ink. Things were even worse in Europe, which long after the war remained beset by severe shortages. In a letter written in June 1947, after her return from England, Kober described the conditions she encountered there:

> *It was quite bearable for a short-term visitor like me. It's different for the people who live there, and have had it for seven years. One of the tutors at St. Hugh's spent a week of her Easter vacation reweaving the elbows of her only decent tweed jacket. Sir John uses the gummed paper of the blank strip on the outside of a page of stamps to cover errors and changes in the ms. he's getting ready to publish. He can't get erasers or ink eradicator. I left him everything I could in the way of writing equipment. He protested very feebly, and ended by saying the things would be "most welcome." All this in one of the countries on the winning side.*

Acutely aware of these conditions, Kober spent the late 1940s sending care packages to overseas colleagues she'd never met, in a one-woman campaign to ameliorate postwar shortages. Her efforts recall those of Helene Hanff, whose epistolary memoir, *84, Charing Cross Road,* recounts her doing likewise for the staff of a London bookshop.

One of Kober's regular beneficiaries was Johannes Sundwall, the Finnish scholar who had dared defy Evans by publishing some of the Knossos inscriptions. Of all the people working on the problem, Sundwall was the one Kober respected most. His approach to the decipherment was of a piece with her own: cautious, scientific, and thorough. He was a good deal older than she; when they began corresponding, in early 1947, she was forty, he about seventy. Though they would never meet, her letters to him—warm, charming, forthcoming, and uncharacteristically girlish—suggest that Sundwall became a cherished intellectual father figure, one of the few people in the world besides John Franklin Daniel by whom she felt truly understood.

Kober had long admired Sundwall's work from afar: It had been impossible to contact Finland during the war. As a result, she was quite unprepared for the letter that arrived at her home in January 1947. "Dear Miss Kober," Sundwall wrote, "I have read your papers on the Knossian tablets and . . . wish to compliment you on your methodical treatment of the most intricate problem that is met with in the Ancient History. . . . I am very glad that you are interested in these problems and hope for collaboration on the [most] interesting and difficult problem of decipherment."

Her reply has the ardor of a schoolgirl who had just received a letter from a favorite film star. "Of all the people in the world," she wrote, "you are the one with whom I have most wanted to correspond. For that very reason, I have been afraid to write. Before the war, it was because I didn't think I had anything to say worthy of your attention, and since then . . . because I didn't have your address. . . . I have carefully laid aside a copy of each article I did write (there are only two so far) to send you when the war was over."

Besides sending Sundwall her articles, Kober would send him many parcels. One, as she wrote in an accompanying letter, contained "coffee in the bean and some soluble coffee called Nescafé"—and here, on the typewritten page, she has carefully drawn in the acute accent on the final *e* by hand—"which is not as good, but lasts longer, although the jar must be kept closed, because it attracts moisture from the air and becomes as hard as stone if one is not careful." On another occasion she mailed him an orange, which she had coated in wax so it could withstand the journey. On a third, she made up a package of Nescafé, sugar, oranges, and rum chocolates, writing, in German, the mutual language in which Sundwall felt most comfortable, "Höffentlich sind Sie nicht Teetotaler (kennen Sie das Wort?)"—"Hopefully you're not a teetotaler (do you know the word?)." After a grateful Sundwall sent her a photograph of himself imbibing one of her gifts, she replied, almost coquettishly: "Mother and I decided it was a picture of you drinking Nescafé, and were delighted when your letter confirmed our guess. Are you sure you haven't made a mistake of about twenty years in your age? After seeing that picture, I am sure you are no more than fifty."

Sundwall's publication of the contraband inscriptions in the 1930s had helped Kober greatly: Without them, she would not have had enough material for even preliminary conclusions. Now, a later book of Sundwall's, *Knossisches in Pylos,* which contained additional inscriptions, would prove a huge boon. It was published in Finland in 1940, but Kober was able to get a copy only in 1947, after searching for more than a year. When it arrived, as she later wrote, she "spent two very happy days . . . copying the inscriptions."

The book also played a role in her rigorous preparation for her trip to England. Because she was under pressure to copy as many inscriptions as possible in her brief time in Oxford, she spent the weeks before her departure training for the task like an athlete preparing for the Olympics. Using the inscriptions in Sundwall's new book as test material, she put herself through rigorous time trials at the dining table. "I've timed myself," she wrote Myres in February 1947, "and think I can copy between 100–125 inscriptions in a twelve-hour day."

What might have made more sense, though, as she acknowledged, was practicing when her fingers were stiff with cold: Myres had warned her that indoor temperatures throughout Britain were almost unbearably low, a consequence of the continuing fuel shortage.

"[Myres] mentions having a 'severe chill' in every letter, and is apparently confined to his room, not only because of illness, but probably also because of the lack of fuel," she wrote John Franklin Daniel in February. "I feel guilty at the thought that I can work in a nice warm house here. I remember only too well what it was like to try and work when the best temperature our fuel ration permitted was 60 or 55."

By March 6, the day before she sailed, Kober had reached a state of hopeful pragmatism. "I'll be content to copy what I can," she wrote to Myres that day. "Ploughing through snow, or wading through rain, March 13 will see me at Oxford, and happy to be there."

On the seventh, armed with writing materials and warm clothing, Kober boarded the *Queen Elizabeth* for the six-day passage to England. She planned to learn Ancient Egyptian on the boat trip over.

* * *

ONE THING KOBER did *not* bring along was the mountain of publications about Linear B that had sprung up in the half century since Evans unearthed the tablets. "Everything that's been written on Minoan is in my files—and much of it is completely worthless," she had written Myres before she left New York. "The useful things I know practically by heart." What she was too modest to add was that the single most vital contribution published so far was a paper of her own—the second of her three major works—which had just appeared in the *American Journal of Archaeology.*

The article was officially published in 1946, although the issue that contained it, delayed by postwar printing problems, did not actually come out until 1947, just before Kober left for England. Titled "Inflection in Linear Class B," it picked up where her first major article, from 1945, had left off.

In the earlier paper, Kober had shown that the Minoans spoke an inflected language. Now came the real payoff from that demonstration: In a discovery that would have enormous implications for the decipherment, she now homed in on precisely what happens when *an inflected language* is written in *a syllabic script.* As a result, the complex interlacements between the Minoan language and the Minoan writing system could start to be untangled.

It was Kober's most dramatic advance so far, and there were three actors in the drama. The first were the stems of Minoan words, analogous to English stems like the noun *kiss-*. The second were the suffixes attached to Minoan words, like *-es* in the plural noun *kisses.* The third was the Linear B character that bridged the gap between stem and suffix. (If English were written syllabically, a single character—representing "se"—would be used to write the last consonant of *kiss* plus the first vowel of *-es* in the inflected word *kisses.*)

Whenever a syllabic script writes an inflected language, these three actors assume a very particular relationship. The relationship is deceptive by nature: It produces a weave that looks hard to unravel but that actually, once only a few sound-values are known, can be picked apart quite easily. In her 1945 paper, Kober had illustrated this relationship with a hypothetical example from Latin, using the verb *facere,* "to make," whose perfect-tense stem, *fec-,* can be combined with a variety of endings:

> Let us suppose, for instance, that Latin was written in a syllabary . . . (with separate signs for the five vowels, and all the other signs representing a consonant and vowel combination), and that we know nothing about either the language or the script used. We find two very similar statements, one ending in the word *fecit* ["he made"], the other in the word *fecerunt* ["they made"]. We would have no way of telling that the words were related, since only the initial sign (*fe*) would be the same for the two words. It would be necessary, before any further progress could be made, **to discover in some way that the signs for *ci* and *ce* had the same consonant**. After that fact had been established, there might be enough material to show that **the sign for *t* and the sign for *runt* alternate with sufficient regularity to permit the supposition that they are inflectional variations**.

In her 1946 paper, Kober expanded decisively on this idea. To read it is to feel the back of one's neck tingle.

As scholars knew, most Minoan words were three or four characters long. Now Kober trained her sights on one pivotal character—usually the third character in a given word. This character linked the stem of a Minoan word with its suffix, just as the *se* of *kisses* does in our imaginary English syllabary. It would come to be known as the "bridging" character, and it would prove crucial in the decipherment.

To show how bridging characters work, Kober looked at the stems of several Minoan nouns. (As she explained, one could reliably spot the nouns of Minoan without being able to read Minoan: Since many of the tablets were inventories, it was a safe bet that each item in a list—"man," "woman," "goat," "chariot"—was a noun. It was equally safe to assume that with rare exceptions, all nouns in a given list shared the same grammatical case.)

As she studied her nouns, Kober spotted eight that had a common ending: -𐀯𐀚. She called this ending Case I. The nouns appearing in Case I included these:

𐀒𐀜𐀯𐀚 and 𐀶𐀪𐀯𐀚

Elsewhere on the tablets, she spotted the same stems with a different ending, -𐀯𐀊. This she called Case II. The words now looked like this:

𐀒𐀜𐀯𐀊 and 𐀶𐀪𐀯𐀊

She found these stems with still another ending, the one-character suffix -𐀰. This she called Case III:

𐀒𐀜𐀰 and 𐀶𐀪𐀰

From these three cases, Kober built a paradigm. Not every noun could be found in every case on the tablets, but she seeded the paradigm with as many examples as she could. Michael Ventris would waggishly name these trios of related forms "Kober's triplets":

	Noun 1	Noun 2	Noun 3	Noun 4
Case I:	𐀶𐀪𐀯𐀍	𐀒𐀜𐀯𐀍	𐀀𐀨𐀯𐀍	—
Case II:	𐀶𐀪𐀯𐀊	𐀒𐀜𐀯𐀊	—	𐀯𐀓𐀯𐀊
Case III:	𐀶𐀪𐀰	𐀒𐀜𐀰	𐀀𐀨𐀰	𐀯𐀓𐀰

Kober homed in on the "spelling change" that affects the third character in each word. In Cases I and II the third character is 𐀯, but in Case III it becomes 𐀰. What, she wondered, accounted for the change?

The changeable nature of this character, Kober realized, signified much. This was the "bridging" character—the link between a word's stem and its suffix. As such, it comprised a piece of each: the last consonant of the stem plus the first vowel of the suffix—a single character that does the work of *se* in ***kisses***. To illustrate this character's role, Kober again turned to a hypothetical example from Latin, this time involving the noun stem *serv-*, "slave." Here *serv-* is shown in three different cases (*servus, servum,* and *servo),* with the hyphen marking the boundary between stem and suffix:

Case I (nominative): serv-**us**	"the slave" (subject)
Case II (accusative): serv-**um**	"the slave" (direct object)
Case III (dative): serv-**o**	"to/for the slave" (indirect object)

Suppose we expand on Kober's example by imagining once more that Latin were written with a syllabic script. The three cases of *serv-* might then be divided into chunks as shown below. It is important to note that when the words are written syllabically, the final *-s* of *servus* is missing, as is the final *-m* of *servum*. There is a reason: Syllabaries like Linear B—in which each character stands for one consonant plus one vowel—are bad at representing final consonants, and often simply delete them.

Here are the three words, written syllabically. This time, the hyphen indicates the boundary between *syllables*:

Case I: ser-**vu**

Case II: ser-**vu**

Case III: ser-**vo**

Our little paradigm contains just three distinct syllables: "ser," "vu," and "vo." We can drive Kober's point home graphically by creating a three-character syllabary with which to write them; I have arbitrarily chosen ❶, ❷, and ❸ as the characters in our tiny syllabic script. The three syllables will now be rendered this way: "ser" = ❶; "vu" = ❷; "vo" = ❸.

Rewritten in the syllabic script, our little paradigm looks like this:

Case I: ❶ ❷

Case II: ❶ ❷

Case III: ❶ ❸

Notice what happens in the switch from alphabet to syllabary. We see, correctly, that the three words share their initial syllable, represented by ❶. But we also see—wrongly—that the second syllable of Cases I and II is identical, written with ❷ each time. Now look at the paradigms side by side:

Written with an Alphabet		Written with a Syllabary			
Case I:	serv-**us**	Case I:	❶❷	=	ser-vu(s)
Case II:	serv-**um**	Case II:	❶❷	=	ser-vu(m)
Case III:	serv-**o**	Case III:	❶ ❸	=	ser-vo

With an alphabet, the difference between *servus* and *servum* is plain. With a syllabary, it is completely obscured: Both are written ❶ ❷.

Our syllabary deceives us in other ways. The alphabet tells us that in all three words, the second syllable starts with the same consonant: "**v**us," "**v**um," "**v**o." The syllabary lies about this fact. Now two different characters, ❷ and ❸, are used to write that syllable, depending on the word's case. This "spelling change" from ❷ to ❸ is crucial: ❷ and ❸ are the "bridging" characters, representing *both* the last consonant of the stem and the first vowel of the suffix. The character changes because the vowel of the suffix ("v**u**s" and "v**u**m" in Cases I and II; "v**o**" in Case III) has changed.

To visualize the role of bridging characters in a "science of graphics," one must mentally split them down the middle, like the contested baby in the King Solomon story, with each "half" claimed by a different syllable:

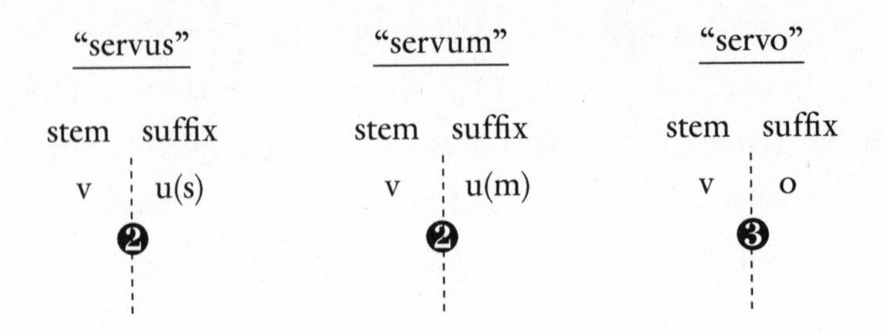

This, Kober realized, was precisely what caused the change in the third syllable of the nouns in her paradigm, repeated here:

	Noun 1	Noun 2	Noun 3	Noun 4
Case I:	𐀀𐀖𐀯𐀍	𐀅𐀄𐀯𐀍	𐀞𐀑𐀯𐀍	—
Case II:	𐀀𐀖𐀯𐀊	𐀅𐀄𐀯𐀊	—	𐀯𐀚𐀯𐀊
Case III:	𐀀𐀖𐀵	𐀅𐀄𐀵	𐀞𐀑𐀵	𐀯𐀚𐀵

It was as though these "bridging" characters, too, had been split down the middle, incorporating the end of the stem and the beginning of the suffix in equal measure. This accounted for the change in spelling from 𐀯 to 𐀵 in Case III:

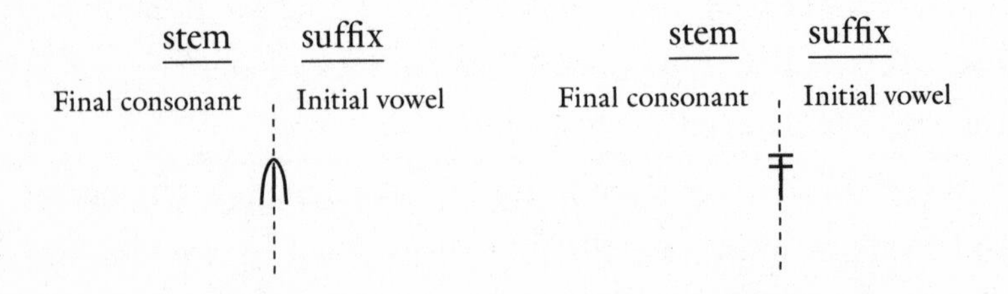

This one-character bridge may look like a small thing. But in isolating its function, Kober had taken an immense step forward. "If this interpretation is correct," she wrote in her 1946 paper, "we have in our hands a means for finding out how some of the signs of the Linear Class B script are related to one another." In the example above, for instance, we can tell instantly that 𐀶 and 𐀵 share a consonant but have different vowels, just as the Latin syllables "vum" and "vo" do.

With a foot in one syllable and another in the next, bridging characters were the linchpins of Minoan words. By identifying and describing them, Kober had found a way of establishing the *relative* relationships among the characters of the script without having to know any of their actual sound-values. And on this linchpin the decipherment would turn, although she would not live to see it.

ON MARCH 13, 1947, when Kober arrived in England, she immediately regretted not having practiced copying with stiffened fingers. It was even colder there than she had expected: She had arrived at the tail end of the brutal winter of 1946–47, famous even now in the annals of British weather. "I've been devoting all the time available to copying Minoan inscriptions, sometimes a more difficult job than you'd think, when the room temperatures hover around forty degrees," she wrote Henry Allen Moe, the secretary of the Guggenheim Foundation, from Oxford in early April.

Having access to so much data was a decidedly mixed blessing. On the one hand, if she could copy it all in her five weeks in England, her store of available inscriptions would increase tenfold. On the other, it would mean that she would have to start her painstaking analytic work all over again. It had taken her five years just to analyze the two hundred inscriptions she already had.

Despite the spartan conditions, Kober reveled in British university life. "I had a most delightful time, staying at St. Hugh's College as a sort of honorary member of the Senior Common Room, and finding out what life at Oxford was really like from the inside," she wrote John Franklin Daniel afterward. "Everybody was so nice, I've come back with a severe case of swollen head."

She, in turn, had nothing but praise for Sir John Myres. "He says what he thinks, firmly, decisively, and oh, so politely," she wrote after her trip. "I can hardly realize I have seen him for only six weeks, once. He is a wonderful man."

Myres let her copy whatever she wanted from Evans's trove of inscriptions, with one proviso: She was to publish nothing based on the material until *Scripta Minoa II* came out. This did not discourage her. The book was due out in early 1948, less than a year away. It would take her at least that long to analyze the welter of new data.

Kober left England on April 17. Before she sailed, she offered to help Myres prepare the manuscript of *Scripta Minoa II* for publication. The sooner the volume was out, she reasoned, the sooner she would be able to make her own discoveries known to the scholarly public.

ON APRIL 25, when the *Queen Elizabeth* docked in New York, it brought Kober home to an ocean of work. "I have so much to do, I hardly am aware that spring is coming," she wrote to Daniel shortly afterward. "My trip to England was successful beyond my wildest dreams—in fact, too successful, since Sir John insists that I go to Crete to check the originals of the Knossos inscriptions for him, and I don't know how the President of Brooklyn College will feel about my running off in the middle of the next semester."

A few months earlier, knowing how much time it would take to analyze the Oxford data, Kober had requested a renewal of her Guggenheim Fellowship. Her letter to Henry Allen Moe, from January 1947, offers a masterly account not only of her approach to the decipherment but also of her willingness to forsake security for science:

"Dear Mr. Moe," she wrote:

> *After a long debate with myself, I finally decided to request the renewal of my Fellowship for another year. The chief reason for my hesitation was that, since Brooklyn College cannot be expected to give me any assistance during a second year of leave, my financial situation will be precarious. It seems to me, however, that at this stage of my work it would certainly be selfish for me to put any personal considerations in the way of a possible successful result. . . .*
>
> *The reason I feel compelled to ask for more time is the unexpected, but very welcome, permission given me by Professor Myres to go to England and see the unpublished Minoan inscriptions.*
>
> *According to report, the total number of inscriptions found by Sir Arthur Evans at Knossos was about 2000. Of these, some 200 have been published. . . . I think (and since my work so far has proceeded according to schedule, my estimates seem to be fairly accurate) that it will take*

about a year to bring the new material into conformity with the old. . . .

Perhaps I had better explain what the process of classification is. Most scholars seem to work with only two files, one in some kind of pseudo-alphabetical order (the system of writing, it must be remembered, is still unknown, and an artificial order has to be set up), the other in reverse, beginning with the end of the word. While these files are essential, they are not enough. One reason why I have been able to do more with the Minoan scripts is because my files contain more. I use, in addition to the two just mentioned

1—a word file, in which enough of the context is set down to show the use of the word

2—two sign-juxtaposition files, one listing every sign used according to the sign that precedes it, the other according to the sign that follows it. These files are the basis for further analysis of possible word roots, suffixes and prefixes.

3—a general file, in which any two signs appearing together in a word are listed, whether they are juxtaposed or not. This file is needed in order to check . . . signs which alternate with one another in certain positions.

Other, more complicated files are built up when the inflection system becomes apparent, but their exact nature cannot be predicted in advance. . . .

Two theories I had in September have now reached the stage of practical certainty. One is that the three different types of Minoan script, Linear Class A, Linear Class B and the Hieroglyphic-Pictographic, seem to record three different languages. The other is that Linear Class B was a strongly

> *inflected language. I have apparently reached the stage where I can predict certain inflectional variations. . . .*
>
> *It seems fairly safe to say that, with the new material, the inflection system, at least of nouns, should become quite apparent. In that case, it should be possible to assign phonetic values to the signs with a reasonable degree of accuracy. Decipherment will then depend on whether the language turns out to belong to a known or unknown linguistic group.*
>
> *If I have another year, I think I can promise that by September, 1948, I will know whether early decipherment is possible. . . . I have reached exactly the right stage in the work to get the maximum benefit from the new inscriptions.*

She closed the letter, "I will be content with whatever decision is made."

The decision reached Kober in England at the end of March 1947: Her application had been denied. The letter from the foundation gave no reason beyond the generic one, that the number of applicants far exceeded the funds available. It seems fair to assume, though, that her constitutional caution, and correspondingly slender record of publication (just the kind of tangible evidence grant-givers like to see), had at least something to do with it.

Characteristically, she made the best of it. "I am in a way relieved, since my financial condition will be better if I go back to work in September," she wrote in a gracious reply to Moe. "With this year of solid work behind me, I should make progress fast." But the rejection was among the first in a series of deep disappointments that would define the last years of her life.

* * *

THE FIRST DISAPPOINTMENT had come four months earlier, at the end of 1946. That November, on the same day she wrote to Myres asking to see the Knossos inscriptions, Kober wrote a similar letter to an American archaeologist, Carl W. Blegen of the University of Cincinnati. Blegen was sitting on another cache of inscriptions altogether: hundreds of clay tablets, inscribed with what looked a great deal like Linear B, that he had unearthed at Pylos, on the Greek mainland, in 1939.

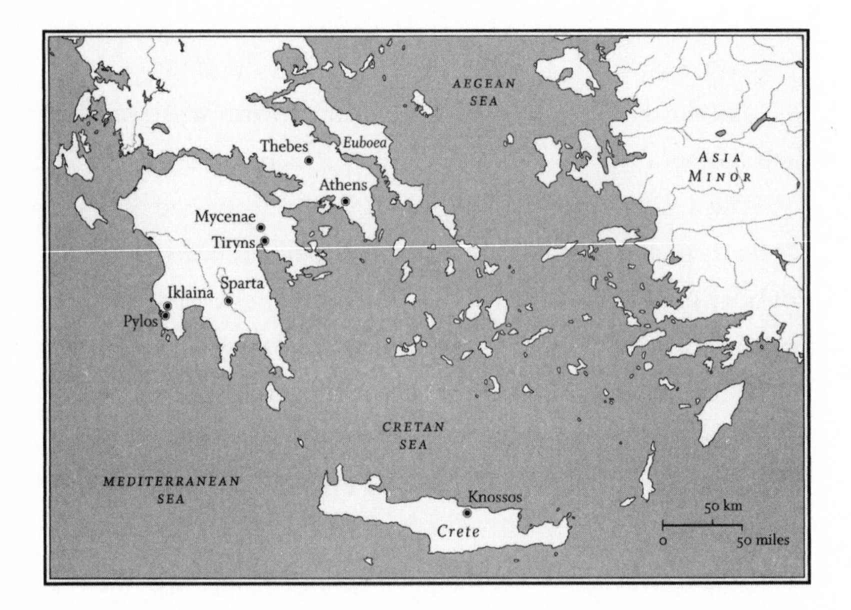

Blegen was even luckier than Evans had been. Where it had taken Evans a week to find his first tablet, it took Blegen less than a day. As the story went, Blegen arrived in Pylos in April 1939 and looked around for a place to dig. He chose a nearby hill and asked its owner if he could excavate there. Permission was granted on one condition: that Blegen's workmen not disturb the ancient olive trees growing on the hillside. Blegen agreed to dig a meandering trench that would spare the trees. As a result, he liked to say afterward, Pallas Athene, the Greek goddess of wisdom, who had given man the olive tree, rewarded him.

Early the next morning, Blegen's Greek workmen began digging the crooked trench. Almost at once, one of the men approached him, holding an object he had lifted from the earth. "*Grammata,*" he said, extending his hand to Blegen. It was the Greek word for writing. In the man's hand was a clay tablet, much like those from Knossos, inscribed with similar symbols. Digging deeper, the crew began to uncover the ruins of a small Mycenaean palace. Blegen named it the Palace of Nestor, after the Homeric hero fabled to have ruled there. Inside the palace was a room filled with tablets.

The tablets had been written in about 1200 B.C.—a good two hundred years after those of Knossos—but the symbols on them looked tantalizingly like Linear B. There were about six hundred tablets, roughly a third as many as Evans had found, but each tended to contain more text. Blegen continued work at Pylos until the fall of 1939. Then, with the outbreak of World War II in Europe, he locked the tablets away in a vault in the Bank of Greece. There they remained, until long after the war.

Blegen's discovery threatened to upend Evans's cherished theory of Minoan supremacy: Though Evans had argued that the mainland was a rude, unlettered outpost of the high Cretan culture, here was evidence that a script—much like Linear B—had been used to record the workings of a *Mycenaean* palace. Worse still for Evans's theory was the fact that the palace at Pylos had clearly been a going concern two centuries after Knossos was burned and the Minoan civilization vanquished.

Evans, by then in his late eighties, did not comment publicly on Blegen's find, though he held fast to his ideal of Minoan domination. By his lights (and so his supporters argued), Pylos was merely a Cretan colony that had adopted the Minoan script and had the good fortune to endure beyond the destruction of Knossos. What was certain was that in the field of Aegean prehistory, Blegen's discovery was the most important since Evans had unearthed the Knossos tablets four decades earlier. As a result, every scholar working on Linear B was itching to see the Pylos inscriptions.

Kober's letter to Blegen was, if anything, more deferential than the one to Myres. But Blegen refused her request. "The difficulties in the way of granting it have arisen not from a dog-in-the-manger attitude on our part," he wrote in reply, "but from considerations of a practical nature. The tablets themselves are still packed away underground in Athens in a bomb-proof vault, where they remained during the war, and when they will be brought out again is wholly uncertain. . . . On this side we have no negatives, but only a single set of photographs which are in use all the time. We likewise have but one complete set of accurate transcriptions, also constantly needed in our own work. Under these circumstances we felt ourselves obliged to refuse similar requests in the past; and the situation has not changed today."

Though access to the Pylos inscriptions would let Michael Ventris crack the code, Kober never got to see more than a tiny handful of them: Blegen's demurral dragged on for years. It was a setback but not a shock. "I am a pessimist," she wrote in 1947. "I prepare for the best, but expect the worst. Usually I am pleasantly surprised."

She had more than enough to do in any case. There were the mountains of data from Oxford to sort and analyze, sign lists and vocabulary lists to draw up, and index cards to be cut, punched, and inked. (She had been unable to copy all two thousand inscriptions, but she did manage to copy most of them, bringing the number at her disposal to almost eighteen hundred.) She had also accepted an invitation from John Franklin Daniel to serve on the editorial board of the *American Journal of Archaeology*—to help him, as he wrote, vet manuscripts on Minoan sent in "by crack-pots." ("I rather consider myself an expert on crack-pots," she replied in acceptance, no doubt contemplating the wild, Polynesian-infused theories of some of her more ardent correspondents.)

Also at Daniel's behest, she would soon start work on her third major paper, which would prove the most important of her career. Kober had already made two major advances: demonstrating that the language of the script was inflected, and identifying the "bridging" characters on which the writing of inflected words hinged. In her third article, she would illuminate a set of vital, long-hidden relationships among the characters of Linear B—the discovery that made Ventris's decipherment possible.

7

THE MATRIX

IN SEPTEMBER 1947, KOBER'S FELLOWSHIP year came to an end, and before long, as she wrote ruefully to John Franklin Daniel, she was back "in academic harness" at Brooklyn College. A few months earlier, contemplating her return, she had written to Sundwall, "It must be quite wonderful when teaching is finally over and one can say to oneself that one has only to learn." (Sundwall was by then retired.) But before the year was out, a joint effort by Sundwall and Daniel would offer her the tantalizing possibility of escape.

At the time, the Linear B tablets remained hidden in Europe, those from Knossos stored in the Heraklion Museum, those from Pylos locked away in Athens. Like many Europeans, Sundwall feared another war on the continent. In the summer of 1947, he wrote to Daniel, suggesting that a safe haven for the tablets be established somewhere in the United States. Daniel seized on the idea at once.

Young (he was in his mid-thirties), brilliant, passionate about Aegean prehistory, and possessed of what appeared to be boundless energy, Daniel seemed precisely the person to realize the plan. "Compared to you a hurricane is just a gentle breeze—except that you're constructive," Kober once wrote him admiringly. Philadelphia was chosen as the tablets' prospective home: In the event of another war, it seemed less likely than New York to be attacked. Daniel set about the delicate diplomatic business of persuading the University of Pennsylvania to create an institute devoted to the study of Cretan scripts, to be known as the Center for Minoan Linguistic Research.

He next set about recruiting Kober as its head. "You are the person in this country who is working most actively in the material," he wrote her in early September 1947. "Would you be interested and available for such a project if it were to become feasible?"

With typical pragmatic pessimism, Kober demurred at first. "I have a job which, while far from ideal in many ways . . . does pay well," she replied. "I don't intend to keep it all my life, but at present the kind of work I'd really like"—pure research—"is open only to men."

In the coming weeks, Daniel wore her down. "Dangling . . . the Institute in front of my nose like that, when all I'm teaching this term is Greek and Roman literature in Translation, and high-school level Vergil, is almost more than I can bear," she wrote him in late September. "I daren't say no, and I daren't say yes."

For Kober, finances were a major concern. At Brooklyn, she wrote, she was paid "a big salary for a woman teacher," more than $6,000 a year—roughly $62,000 today. "I know enough about academic salaries for women to realize that financially I can't do better, at least for another ten years," she told Daniel. "All the same, if it weren't that I have Mother to consider, that wouldn't make any difference."

But over the next few months, as Daniel continued his backstage machinations, the job began to tug at her increasingly. By chance Roland Kent, a well-known professor of Indo-European linguistics, had just retired from Penn, and Daniel hit upon his departure as a way of bringing Kober to the university, with her duties divided between running the Minoan institute and teaching classes on subjects like Greek phonology, a topic much dearer to her heart than high-school-level Vergil.

"Don't count on it too much, because I am still pretty young in university politics and it may well be that my word will carry even less weight than I timidly think," Daniel cautioned her by mail that fall. "Furthermore I am afraid that there is something of a prejudice against giving faculty appointments to women. How strong it is I don't know, but will soon find out."

After several months, Daniel was able to report that his efforts at Penn were going slowly but well. If he could arrange it, he wrote Kober in December, he would make a course on Cretan scripts part of her portfolio there. With that, she was hooked. "Your latest letter . . . has me sitting here with my tongue hanging out, and that's bad," she replied. She continued:

> *I'm trying to preserve my equilibrium so that I'll be happy no matter how things turn out. You're heartless. . . . Now you add a course in Minoan scripts!!! I've been looking at the list of courses, and feel much encouraged. I could begin teaching most of them to-morrow, and plan the entire course in a week or two. . . . The salary doesn't concern me too much. . . . The most important consideration is that I'll love the work. . . . If you can do half as well in selling me to*

> *Pennsylvania as you have done in selling Pennsylvania to me—it's in the bag.*

As Kober and Daniel both knew, the gears of university administration grind with the speed of geologic time, and there would be no word on the job right away. In the meantime, though, he had conceived something equally important to occupy her.

It is hard to imagine now, but among their other effects, the shortages of the postwar years made the dissemination of ideas among scholars a real challenge: The process depended crucially on printing, which remained an uncertain proposition, and paper, which remained a scarce commodity. In the late 1940s many Europeans still had scant access to American academic journals, and vice versa. In the autumn of 1945, Kober lamented that the Brooklyn College library's most recent copy of *Archiv Orientální,* an important Czech journal, was from 1938, "as might be expected." In 1946, she had to enlist the aid of the Czech Consulate to obtain a copy of Hrozný's "decipherment" of Linear B.

The number of people working seriously on the Cretan scripts was so tiny, their mutual contact so limited, and their collective knowledge so atomized, that most were effectively laboring in isolation. Beyond this small group was a larger one of "non-Minoan" archaeologists and classicists, who knew far less about the scripts than they might. What was needed, Daniel realized, was one overarching article, disseminated as widely as possible, that laid out precisely the current state of knowledge about the scripts. He knew just the person for the job, and in early September 1947, he wrote to her, soliciting the piece for the *American Journal of Archaeology*:

> *Everybody is interested in the Minoan script and it is astounding how few people know anything about it. . . . I think it would be useful not only to summarize the most recent work in the matter but to state the problem generally, indicate the various lines which have been followed in attempting to crack the script and saying a brief word about the real merits of the different methods; then, I think, a brief statement, in which I hope you will not be modest, of the present state of the study, with your ideas as to its probable future development . . . [and a summary of] the fallacious attempts to decipher it.*

Though Daniel and Kober's friendship was conducted almost entirely by mail (they met only a few times, usually at scholarly meetings), there existed between them an intellectual sympathy so deep it bordered on telepathy: As it happened, she was just then thinking of writing a state-of-the-field article herself. "One of the remarkable things about you is that you always anticipate what I'm going to say," she wrote by return mail. "I think your suggestion about an article on what is known about Minoan just suits the situation. It will also make the Guggenheim Foundation happy, since it will be a sort of summary of my year's work."

By late September, she had embarked on the torturous process of drafting the paper. "About the article—I hope it comes up to expectations, but I'm beginning to have my doubts," she wrote to Daniel. "My typewriter seems to have taken the bit into its mouth (how's that for a mixed metaphor). At any rate, what has come out isn't at all what I expected." By early October, she was hard at work on the fourth draft; by the middle of the month, as she wrote to Daniel, she had completed "the sixth draft (durn it!)." Throughout the article, she was careful not to cite any inscriptions from the still-unpublished Knossos tablets, honoring her promise to Myres.

In late October, Kober sent Daniel the finished manuscript. "This is the best I can do," she wrote. "I should have rewritten one or two pages, but I'm afraid to touch the typescript, for fear that I will feel like writing another draft. I'm sick and tired of it, truth to tell. I've worked harder on this than on anything I've ever written, and doubt that it's worth beans. If you don't think it's worth printing—well, I agree." In fact, the paper would be the most important of her career, and the one that furnished the most significant step forward in the decipherment thus far.

IN MID-DECEMBER 1947, before her article appeared, Kober made what she called a "slight discovery." Sifting her cards, she had managed to pinpoint the special function of the character ⊜, colloquially known to investigators as "button." Though the discovery was small, the noteworthy thing about it is that it was she, and not Michael Ventris, who made it: Until now, historians have attributed it to Ventris, who reached the same conclusion independently more than three years later, in January 1951.

What first Kober and then Ventris noticed was that ⊜ kept cropping up in a very particular spot: at the ends of words. As Kober knew from perusing the tablets' inventories, many of those words were nouns. In addition, each "buttoned" noun typically followed another noun, resulting in this characteristic sequence:

noun | noun-⊜.

Though she still had no idea how ⊜ was pronounced, Kober realized that in the language of the tablets, it must be the conjunction "and." The symbol was functioning as a kind of suffix: Tacked on to the end of one noun, it linked that noun to the one before it.

As Kober knew, "and" can function as a suffix (or prefix) in other languages. In Latin, "and" (*-que*) can be tacked on to the end of a noun, linking it to the previous one. (A well-known example is the phrase *Senatus Populusque Romanus*—"the senate **and** the people of Rome"—abbreviated SPQR and ubiquitous on Roman coins, documents, and the like.) In Hebrew, "and" (*v-*) attaches to the *beginning* of a noun, linking it to the noun before: *tohu vabohu,* "without form **and** void," from the book of Genesis.

Kober concluded—correctly—that ⊜ worked similarly in the Minoan language. But she was never able to make her finding public, and as a result never got credit for it. This was partly an accident of timing: She made the discovery on December 14, 1947, too late to prepare a paper on the subject for the year-end annual meeting of the Archaeological Institute of America, as she would have liked. "Too bad I didn't discover it a couple of weeks ago," she wrote Daniel regretfully.

She may not have had time in any case. By the late 1940s she had become enmeshed in a gargantuan, long-distance effort to proofread, fact-check, and type Myres's work on *Scripta Minoa II.* Though to Kober, as Professor Thomas Palaima points out, the job had the quality of a sacred duty—one that might let her ensure the absolute accuracy of the material Myres had inherited from Evans—it was proving to be a frustrating, unrewarding, and labor-intensive enterprise that at bottom was little more than secretarial work. And she was carrying a full load at Brooklyn College all the while.

Even had Kober been able to make her findings on ⊜ public, it is quite possible that Ventris would not have learned of them anyway. Her narrative and his don't truly begin to converge until the spring of 1948, when he started corresponding with Linear B investigators around the globe. From the letters between them, which begin that March, it is plain that Kober's work remained unavailable in Britain. In a letter to her in May 1948, Ventris acknowledges receiving copies of her articles—"I'm extremely glad to have them," he wrote—in a manner that strongly suggests he had not seen them before.

Kober and Ventris would meet only once, briefly, in England, in the summer of 1948, and there is no direct record of what either thought of the other. But it is clear from the absence of the kind of warm, respectful correspondence that Kober maintained with colleagues like Daniel and Sundwall that she did not hold Ventris in especially high esteem.

She had no use for amateurs, many of whom persisted in writing her with their latest theories on the Cretan scripts. One of her most chronic correspondents—whose letters are charming to read today, but to Kober were undoubtedly a source of immense irritation—was William T. M. Forbes, a Cornell University entomologist and dabbler in ancient scripts. (It was he who thought the Minoan language was a form of Polynesian.) His letters to Kober, written over a period of years, contain page upon page of unsupported linguistic speculation before closing with cheery sign-offs like "But now back to the Lepidoptera for a time." Though Kober's replies have not been preserved, it is evident from Forbes's letters that she took the time to write him back, at pedagogical length, which left him grateful if unpersuaded. Her handwritten notes in the margins of his letters to her ("No!" "Right!!") attest to their claim on her time.

To Kober, Ventris appeared to be yet another hobbyist awash in wild, unfounded enthusiasms. Neither linguist nor archaeologist, he was an architect who worked on the tablets in his spare time. Their correspondence had an inauspicious beginning: His first known letter to her, on March 26, 1948, contains a slew of things that must surely have set her teeth on edge. "I should be very interested to hear how far you have got at present, and particularly if you have any ideas on phonetic values," Ventris wrote. He restated his conviction that the Minoan language was a form of Etruscan, a belief he had first made public in a 1940 article in the *American Journal of Archaeology,* published when he was a teenager. Neither his talk of sound-values nor his identification of a specific language as Minoan would have sat well with Kober, who remained resolutely agnostic on both subjects to the end of her life.

"Minoan is only a part-time job with me," Ventris wrote in the same letter, "and there is always the problem of getting enough material to work on. In fact, I don't feel like coming out publicly with any more theories until I've laid my hands on all the available material." He went on to describe setting up a network of Linear B characters that had consonants or vowels in common—relationships Kober had already described in her influential paper of 1946, "Inflection in Linear Class B." He also talked of setting up a "grid" by which those shared consonants and vowels could be plotted, which was precisely what

Kober had done by then in the state-of-the-field paper Daniel had commissioned.

Later in the spring of 1948, on the verge of quitting his day job to immerse himself completely in Linear B, Ventris wrote her, "At the moment I'm engaged in a rather tricky architectural competition, but after that I think I shall have to devote the rest of the year whole time to Minoan, because it isn't really worth doing in fits and starts," a line that to Kober must have fairly screamed "rich dilettante."

It wasn't that Kober saw the decipherment as a competition. "Rivalry has no place in true scholarship," she later wrote to Emmett L. Bennett Jr., a young classicist at Yale. "We are co-workers. I will be glad to help you in any way possible. The important thing is the solution of the problem, not who solves it." Nor would she have felt threatened by amateurs like Ventris in any case. It was that she could abide neither their unfounded guesswork nor their persistent correspondence, which, through a combination of good manners and fealty to the subject at hand, she spent too much of her scarce free time answering.

BY THE END of 1947, Daniel, continuing his campaign to bring Kober to Penn, had reason to feel encouraged. "Please send me FAST a complete list of your publications, including book reviews, classes you have taught, and the titles of lectures you have given," he wrote her in early December. "I have just come from a meeting of the committee to appoint a successor to Kent, and . . . there was some very strong support for your nomination. . . . What this all adds up to is that I think there is an excellent chance."

At Daniel's request, Kober solicited letters of recommendation from some of the eminent scholars who had taught her over the years. One, Franklin Edgerton, whose Sanskrit classes she attended at Yale, was happy to comply, if amazed. As he wrote her:

> *I am indeed astonished by the news conveyed in your letter. . . . The astonishment is not uncomplimentary to you; it is chiefly on account of the fact that I had no idea that the University of Pennsylvania ever had appointed or would appoint a woman to a major position. If they actually do offer you a full professorship at a good salary . . . it seems to me that I would take it, if I were you.*

By this time, Kober needed no persuading. "Everybody tells me this job, if it materializes, would be a wonderful opportunity," she wrote to Daniel. "I know that too." She continued:

> *It's too good to be true. I write letters and . . . talk about it, but I don't believe it. All the same, I've been figuring out what I would do if the job actually materializes. I think I would work it this way: I'd take it, and try to get a year's leave without pay from Brooklyn College. In that way, I can find out how it really works out, and if I'm not as good as I think I am, can retire gracefully at the end of the year, to everybody's relief, and come back here. If it works out, I can resign here. . . .*
>
> *I'll have to figure on doing very little with Minoan that first year—but getting [the research center] into order and plugging away at new courses will be a very valuable*

> *experience, and Minoan is at the stage where any new information may be the clue. . . .*

She closed the letter: "I've never told you I'm grateful for all you're doing. I am. It's been fun, too."

Amid all this, Kober was attempting to arrange two overseas trips. The first was a return to England: Sir John Myres wanted her to help lobby the Clarendon Press, an imprint of Oxford University Press, to speed the publication of *Scripta Minoa II,* which had ground to a halt amid postwar retrenchments. In the spring of 1948, she finally secured passage—no small trick, given the austerities still affecting ocean travel then. "First Class. Ouch!" Kober wrote Daniel in late April. "But that was the only way I could go." She would sail from New York on the RMS *Mauretania* on July 21 and reach Oxford by the end of the month. She would sail home from Liverpool on September 10, arriving in Brooklyn just before school began.

The second trip was to Crete, where Myres wanted her to go to check Evans's transcriptions against the tablets in the Heraklion Museum, provided they were accessible by the time she got there. But even if they were, she had neither the time nor the money to make the journey: Brooklyn College would grant her only six months' *unpaid* leave. "Six months is ample time for me to starve to death, since I've spent all my money last year," she had written to Daniel in late 1947. She added—and it is wrenching to read in retrospect—"Six months mean very little to me, but of course, they mean a lot to Myres." (Myres would outlive her by four years.)

"My only solution," Kober wrote, "would be to stop all scholarly work for six months, write a detective story and hope it would be a best-seller." Of course, her body of work, read in sequence, *was* a detective story, and a great one, though it would take Michael Ventris building on the foundation she erected to solve the mystery. In the end, Kober's worries about the Cretan trip were moot. The Knossos tablets remained in hiding for years, and she would never get to see them.

IN EARLY 1948, Kober's state-of-the-field article appeared in the *American Journal of Archaeology* under the crisp title "The Minoan Scripts: Fact and Theory." Erudite, authoritative, and coolly polemical, it began:

> The basic distinction between fact and theory is clear enough. A fact is a reality, an actuality, something that exists; a theory states that something might be, or could be, or should be. . . . In dealing with the past we are concerned, not with something that exists, but with something that has existed. Our facts are limited to those things [from] the past which still exist; everything else is theory, which may range all the way from practical certainty to utter impossibility, depending on its relationship to known facts.

The heart of the paper picked up where her 1946 article had left off, with the critical third character—the "bridging" character—in Linear B words. From that character, Kober had long known, sprang a network of relationships among the sounds of the Cretan language. Now, in this paper, she began to plot those relationships by means of a grid.

As before, Kober paid particular attention to what happens

when an inflected language bumps up against a syllabic script. By this time she had discovered additional words that seemed to fit the pattern of inflection she had previously described. In the new paper, her expanded paradigm contains six words (she believed they were all nouns) in each of three cases—six sets of "triplets," as Ventris would call them. As it happened, the three words in Column B (𐀒𐀜𐀯𐀍, 𐀒𐀜𐀯𐀊, and 𐀒𐀜𐀰) would play a critical role in the decipherment.

	A	B	C	D	E	F
Case I:	[Linear B]	𐀒𐀜𐀯𐀍	[Linear B]	[Linear B]	[Linear B]	[Linear B]
Case II:	[Linear B]	𐀒𐀜𐀯𐀊	[Linear B]	[Linear B]	[Linear B]	[Linear B]
Case III:	[Linear B]	𐀒𐀜𐀰	[Linear B]	[Linear B]	[Linear B]	[Linear B]

Once again, Kober homed in on the third sign of each word (or, for shorter words, like those in Columns D and E, the second sign). This was the "bridging" character, whose function she had illustrated so elegantly in her previous article. It is this character that is at the heart of the matrix, or grid, that she unveiled in the new paper. The grid is modest—just five signs by two signs—but it presents, for the first time anywhere, some *relative* sound-values for characters in the script. Kober did not invent the concept of the grid, nor did she claim to: Its use in archaeological decipherment dates to the seventeenth century. But her great innovation, as Maurice Pope writes in *The Story of Archaeological Decipherment,* "was the idea of constructing such a grid *in the abstract* . . . without settling what particular consonant or vowel it might be that a particular set of signs had in common."

Captioned "Beginning of a Tentative Phonetic Pattern," Kober's grid looked like this:

	Vowel 1	**Vowel 2**
Consonant		
1	[symbol]	[symbol]
2	[symbol]	[symbol]
3	[symbol]	[symbol]
4	[symbol]	[symbol]
5	[symbol]	[symbol]

Each symbol in the grid is one of Kober's "bridging" characters, and each character's position marks, so to speak, its phonetic coordinates. Reading across Row 1, for instance, we see that [symbol] and [symbol] start with the same consonant but end in different vowels—whatever those consonants and vowels might be. Reading down Column 1 tells us that [symbol], [symbol], [symbol], [symbol], and [symbol] start with different consonants but end in the same vowel. Though the specific sound-values remained unknown, Kober's grid made it possible to show the *relative* relationships among these ten characters. A comparable grid for English—and here the sound-values have been assigned arbitrarily—might look like this:

	Vowel 1	**Vowel 2**
Consonant		
1	ba	be
2	da	de
3	fa	fe
4	ka	ke
5	ma	me

Kober's grid illustrates the web of contingencies that emerges when an analyst plots the changes in the "bridging" character across different cases of the same word. To modern eyes, her grid has the quality of a Sudoku puzzle, in which the interdependencies among its cells ("If I've already used a 5 in this square, I can't use a 5 in an adjacent square") help the investigator arrive at the only logically possible solution. For, as she clearly knew, once the sound-values of just a few Linear B characters were discovered, the entire grid, in explosive chain reaction, would start to fill itself in.

What Kober showed is that when an inflected language is written with a syllabic script, mapping the inflection pattern is a way to "force out" hidden information about the relationships among the signs. It was this information, so masterfully presented in her new article of 1948, that furnished the first viable key to the mapping between sound and symbol in the lost Cretan language.

Near the close of the article, she wrote:

> People often say, in connection with the Minoan scripts, that an unknown language written in an unknown script cannot be deciphered. They are putting the situation optimistically. We are dealing with three unknowns: language, script and *meaning*. A bilingual inscription is useful because it gives meaning to an otherwise meaningless combination of symbols. Those who deplore the fact that no Minoan bilingual has been found, forget that a bilingual is no guarantee of immediate decipherment. The Rosetta Stone was found in 1799. Champollion began his intensive work in 1814, but it was not till 1824, a quarter of a century after the Rosetta Stone was discovered, that he was able to

publish convincing proof that he had found the clue to the decipherment of Egyptian.

"Let us face the facts," she wrote in conclusion:

> An unknown language, written in an unknown script cannot be deciphered, bilingual or no bilingual. It is our task to find out what the language was, or what the phonetic values of the signs were, and so remove one of the unknowns. Forty years of attempts to decipher Minoan by guessing at one or the other, or both, have proved that such a procedure is useless. . . . The people of ancient Crete did not live in a vacuum, nor did they disappear suddenly and completely. They left traces of their languages behind. These traces are no good to us now, because we do not know enough about their scripts to use them intelligently.
>
> The task before us is to analyze these scripts thoroughly, honestly, and without prejudice. . . .
>
> When we have the facts, certain conclusions will be almost inevitable. Until we have them, no conclusions are possible.

IN LATE DECEMBER of 1947, Kober had received a special-delivery letter from Daniel warning of a "slight setback" regarding the job at Penn. "You are no. 2 on nearly everyone's list," he wrote:

> *Your Minoan accomplishments . . . have already been made clear to the Board. . . . What is needed and needed badly is a*

strong endorsement of your ability to handle Indo-European and particularly Greek and Latin linguistics. Several people have written to say that they have no doubt that you could give satisfactory teaching in this field, but several have intimated that you might have to bone up on it. What is needed is a categorical statement to the effect that you are fully qualified in these fields. There is no doubt whatever in my mind that you are, but it is going to take more than my statement to swing this matter. If you can get one such statement from a first-rater, I think that your chances of landing the job will be very good indeed.

In reply, Kober asked him point-blank: "Don't you think a lot of the opposition is really based on the fact that I'm a woman? Even if it isn't specifically mentioned." She went on: "As for the 'boning up'—anybody must 'bone up' for a graduate course, or for a course one has never taught. All my great teachers . . . do their homework, even for courses they've taught over and over again. Of course I'll have to work at preparation."

Daniel answered, "The fact that you are a woman has absolutely nothing to do with the case." The trouble, he wrote, lay with several members of the search committee, Indo-European scholars at the university. One was "weak, lazy, and impressionable"; the other "has the first two qualities, but has a stubbornness which [the first man] seems to lack." A third "has been an assistant professor for seven years, and . . . told me perfectly frankly that he felt that he had to oppose any appointment at a higher rank than that, because it would shut him out forever. Isn't that nice?"

Of the committee members as a group, Daniel wrote,

"I am not being unduly bitter when I say that their controlling criterion seems to be mediocrity. They are third raters themselves . . . and simply do not want to get people here who will show them up. . . . While he has not said so in so many words, I am sure that Crosby, who is chairman of the department, has made up his mind that he will block you if he possibly can. . . . I certainly have learned a lot about human nature, the genus professoricus [*sic*], in this matter!" The university's intransigence continued through the start of 1948, and in February Daniel wrote her, "I am limiting my activities to trying to undermine your rivals."

Finally, in May 1948, Penn made up its mind. "I have bad news," Daniel wrote. The university had appointed the eminent Indo-European linguist Henry Hoenigswald to the post. "I am terribly disappointed about it, perhaps more so than you will be," Daniel continued. "I was dreaming wonderful dreams of the terrific set-up we would have here with you and the Minoan collection. . . . Hoenigswald is a good man, but it won't be quite the same. It may give you some satisfaction to know that you were very close to getting it; if one person had swung from opposition . . . to support, I think that we could have swung it. But that is spilled milk."

"Well, it was fun while it lasted," Kober replied. "I can't say your news was unexpected, because I am a pessimist from 'way back."

There was one silver lining: Penn wanted to go ahead with the Minoan center anyway, with Kober spending one weekend a month in Philadelphia to tend to it. In anticipation, she was named a research associate at the university museum, an honorary title with no stipend.

By mail, she and Daniel began to draw up their plans. Kober compiled a list of people she would invite to be associates of the center—forming a "mutual aid society," as she called it, to share the fruits of their research. "If it works as we hope," she wrote to Henry Allen Moe in July, "scholars of a dozen different countries, now working more or less in isolation, will be able to cooperate, and perhaps our united efforts will solve the problem." At the top of the guest list she put Sundwall and Myres, followed by a few other scholars. Next-to-last she put Bedřich Hrozný, the object of her frequent scorn, whose name she followed with three question marks. In last place was Michael Ventris, followed by four question marks.

BY THIS TIME, Ventris, too, had been enlisted by Myres to help prepare the Knossos inscriptions for publication. The two men had begun corresponding in 1942, when Myres wrote to Ventris to compliment him on his 1940 article in the *American Journal of Archaeology.* At war's end, he wrote to Ventris again, with an offer similar to the one he would make to Kober: permission to see Evans's unpublished transcriptions.

Ventris accepted eagerly, though he would not be able to take up the offer until 1946. Soon afterward, at Myres's behest, he was carefully hand-copying hundreds of inscriptions from Evans's notes and photographs so that they could be reproduced clearly, as Kober was also doing. (With a good two thousand inscriptions, there was ample copying work for them both.) Ventris began spending each night at an elegant table in his London home, drawing Linear B symbols in his impeccable architectural hand.

Where Kober's Linear B transcriptions are serviceable, Ventris's are the work of an artist. Even his everyday handwriting, the letters crisply squared off and perfectly proportioned, the lines absolutely horizontal, is so remarkable-looking that it resembles the penmanship of a skilled blind person writing with the aid of a straight-edge.

19 January 54

Dear Sir John,

Thank you for your letter: I have already tentatively fixed to talk at the British Association on September 6: if your suggestion of a wider Minoan discussion takes place, it will be extremely interesting.

There is a slightly "rival concern" in the Congress for Classical Studies in Copenhagen who are having a 'Minoan day' on August 26. Chadwick or I or both have promised to attend; & they are asking Gelb & Sittig on the decipherment-technique side.

Yours, Michael Ventris

Ventris's everyday handwriting, remarkable in appearance.

"Mr. Ventris would have no trouble getting a job as scribe for King Minos," Kober wrote after seeing a batch of his copying, a remark that appears at once complimentary and belittling. Ventris might be a superb draftsman, she seemed to be saying, but he was no more than that. What is clear, with hindsight, of both Kober and Ventris is that each underestimated the other deeply.

IN MIDSUMMER 1948, plans for the Minoan center were put on hold temporarily, until both Kober and Daniel returned from overseas trips—hers to Oxford and his a long voyage through Greece, Cyprus, and Turkey. Daniel's trip, to scout sites for future excavations by the university museum, would keep him abroad from September 1948 through February 1949; he was due to set sail for Athens on September 10, the day Kober sailed home from England to New York.

On July 21, 1948, Kober embarked for England once more. She would not only need to rouse the Clarendon Press to action but, should they decide to go ahead, she would also have to continue helping Myres prepare the complex, unwieldy manuscript for submission. So much depended on its being published. "Until it is," she wrote, "it will be impossible to make progress."

In Oxford, Kober found Myres even frailer than before, and the state of his health made access difficult. "Lady Myres keeps him in bed Mondays and Wednesdays," she wrote to Daniel from Oxford in August. She was just then in the midst of revising Myres's vocabulary list, which, she wrote, "is in perfect chaos. . . . Every other word requires correction." Perhaps that was not surprising—Myres was an archaeologist and not a linguist—but the state of his data did Kober no favors.

For the first time since her association with him began, Kober expressed reservations about the quality of Myres's work. "I only hope he accepts my corrections," she wrote to Daniel. "I'd hate to have him publish what he has and mention me in connection with it. He is far from well, which makes it difficult to press a point."

Kober understood that Myres was living only for the volume's publication. But her correspondence from this trip makes clear that dealing with him was a struggle from the moment she arrived. "I've had enough trouble getting him to correct the Linear B," she wrote in another letter to Daniel that August. "I've just about rewritten Scripta Minoa II and have the Clarendon Press ready to print. Now I'll have to get Sir John to give up the manuscript to be printed. Better burn this letter." She signed off: "Just now I'm writing out the vocabulary for the Press—a month's work, and less than a week to do it. What a life!"

It is here, in Oxford, that Kober's and Ventris's stories dovetail for the first brief, painful time. Ventris had also been summoned to Oxford in the summer of 1948 to help prepare *Scripta Minoa II* for publication, and he arrived during Kober's visit. But, apparently cowed by the combined scholarly firepower of Kober and Myres, he quickly fled the scene, a pattern of abdication he would repeat throughout his life.

Besides correcting Myres's Linear B vocabulary lists for publication, Kober had a spate of purely scribal duties, including the tedious hand-copying of hundreds of inscriptions for the printer. She also agreed, in retrospect unwisely, to help Myres prepare the manuscript of yet another volume, *Scripta Minoa III*, which would be devoted to Linear A.

There were a few consolations. Foremost was Kober's hope that *Scripta Minoa II* would finally see publication, freeing her to use the inscriptions in her own work. There was the joy of staying again at St. Hugh's College, whose intellectual climate she had pined for in the year since she'd been there. There was also the letter that Daniel mailed to her in Oxford, written days before he sailed for Greece. "Two deans . . . still wish there were some way of getting you a full time appointment at Penn," he wrote. "Who knows: it may work out yet."

Despite her efforts, *Scripta Minoa II* was no closer to publication when Kober sailed for New York in September. Apart from the deplorable state of the manuscript itself was the fact that she and Myres disagreed on a fundamental organizing principle: how to classify and present the nearly two thousand Knossos inscriptions. He wanted to present them much as Evans had, organized by the location in the palace in which they were found. In her own work, by contrast, Kober had spent years grouping the inscriptions together by subject: all the tablets about grain discussed together; the ones about wine; about men, horses, and so forth. To her, this was the more meaningful classification, as it spoke to the tablets' focus on lists, which in turn let her pluck out the nouns and their various inflected forms.

Myres came around to her way of doing things, but a huge sticking point remained: He didn't think the language of Linear B was inflected. "He wants to use my classification of the inscriptions but still insists there are no cases!" Kober had written Daniel in frustration from Oxford. "My classification depends on case."

From this point on, the tone of her correspondence about Myres, once worshipful ("I am really very much in awe of that great historian, J.L. Myres," she had written to Myres himself in 1946), starts to change. In a letter to Sundwall written shortly after her second trip, Kober said that Myres "still thinks . . . that there are no cases, also, that anything he doesn't understand is due to the fault of the Minoan scribe." To Emmett Bennett of Yale, she wrote: "Somewhat against my better judgement, I permitted Sir John Myres to use my classification as a basis for his discussion of the Linear B inscriptions. I didn't like the idea because I would prefer to explain my classification myself, since he still thinks all words are nouns, and all in the nominative case. But I finally said he could do so, provided he stated his interpretation was his own, and not mine. . . . I now plan to publish the classification with my own interpretation as soon as SM is out."

ON SEPTEMBER 18, 1948, Kober arrived in Brooklyn to the start of a new term. Soon she was swamped. "I've been home almost a month now, and haven't done as much as I could do in a couple of days of uninterrupted work," she wrote Myres in October. "It is annoying to have to stop what I'm doing at 11 at night, often right in the middle of something, get ready for bed so that I can get up for school in time, then find that committee meetings, and unexpected visitors, keep me from continuing for a couple of days."

She had not only her school duties to occupy her time but also Myres's typescript of *Scripta Minoa II* to correct for the printer, as well as his handwritten manuscript of *Scripta Minoa III* to begin typing. All this left her barely an hour a day for her own work on Linear B. She could look ahead with pleasure, at least, to Daniel's return from overseas in five months, when they could discuss plans for the Minoan center and even, just possibly, a faculty appointment at Penn. "Wishing you the best of luck and looking forward to seeing you again in February," he had written her in early September, before he set sail.

On December 18, 1948, in Ankara, Turkey, John Franklin Daniel died of a heart attack at the age of thirty-eight.

8

"HURRY UP AND DECIPHER THE THING!"

THE NEWS CAME AS A terrible shock, as you can imagine," Kober wrote Myres in late December 1948. "At present I feel depressed. Daniel was a friend of whom I was fond, as well as the person on whom practically all of my Minoan plans for the immediate future depend. I don't know what's going to happen. . . ." A letter she wrote to Sundwall in March 1949 betrays the increasingly darkening tone of her correspondence. "I did not write more about Daniel because I myself had no information," she said. "He was taken to Cyprus, and buried at Episcopi. That's all I know. It is very sad. I still can't believe it." She added, in an aside of a sort rarely seen in her letters: "I am afraid our civilization is doomed. Whatever happens, the freedom of the individual will be lost, and for generations we will—." The rest of the letter has not survived.

Through all this, Kober continued her work. She had her classes to teach and her papers to grade; she also had the hundreds of inscriptions she had fiendishly transcribed on her second trip to Oxford, which she was now sorting, cataloguing, and copying neatly for reproduction by the Clarendon Press. At this point, she was dealing with about 2,300 inscriptions, more than ten times as many as when she began.

Myres was still sending her batches of the manuscript of *Scripta Minoa III,* the volume on Linear A, written in his nightmarish hand. "I'm neck-deep in Sir John's manuscript about Linear Class A, which I'm supposed to be editing for him," Kober wrote to Emmett Bennett toward the end of 1948. "First, though, I have to type it out—it's manuscript, *manu scriptum.* And in typing, I find he makes errors—so I have to correct them. Considering that I think Linear A isn't going to be much help for decipherment it's a complete waste of time, for me."

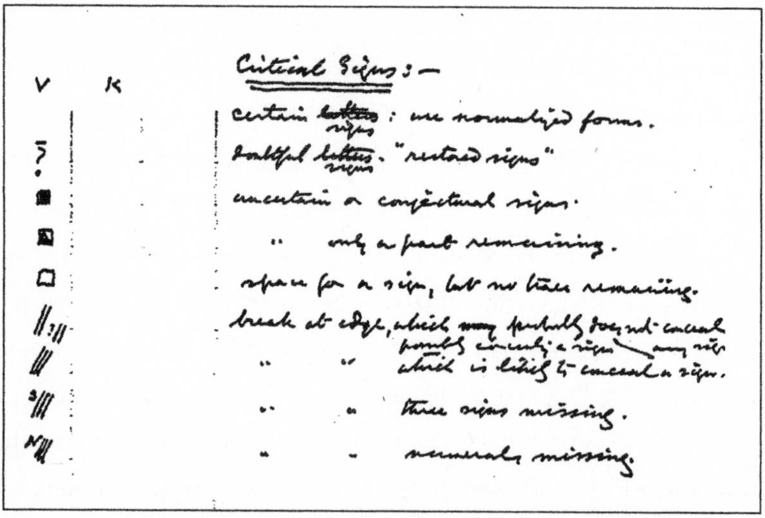

A sample of John Myres's handwriting from 1948.

What is haunting about Kober's correspondence from this period is how often she invokes time: the passage of time, the wasting of time, and the spate of claims on her time. Though she had not yet experienced the first signs of her illness, the very notion of time—and in particular the lack of it—permeates her letters from this point onward.

There were a few bits of good news. One came from the University of Pennsylvania, which in the wake of Daniel's death wanted to go ahead with the Center for Minoan Linguistic Research anyway. With Daniel's successor, Rodney Young, Kober resumed organizing the center, which she planned to have open to scholars by the end of 1949.

Another concerned the Pylos inscriptions. From the time they were unearthed on the mainland in 1939 through the end of the 1940s, only seven inscriptions had been published. Hungry scholars like Kober could do little more than guess at the tablets' contents and steal rare glimpses whenever they could. (In 1942, after attending a lecture by Carl Blegen, the archaeologist who had unearthed the tablets but declined to share them, Kober wrote, "He showed slides of about three inscriptions I had never seen before, so I considered it a perfect evening.")

The Pylos script looked similar to the one from Knossos, but there were visible differences. Each had characters not found in the other: The symbols [symbol], [symbol], and [symbol], for instance, appeared at Pylos but not at Knossos; the symbol [symbol] appeared at Knossos but not at Pylos. Without full access to the mainland inscriptions, Kober would never be able to determine to her own satisfaction whether the scripts were the same. But Blegen continued his refusal. By the late 1940s, Kober had resolved to go quietly to work on his disciple, Emmett Bennett of Yale. "Dr. Bennett . . . is a very agreeable young man whenever he can be," she wrote to Sundwall in 1948, in a barely veiled swipe at Blegen.

The cautious correspondence between Kober and Bennett had begun in June 1948, when Bennett, who had done his doctoral dissertation on the Pylos tablets, wrote to her with a recommendation. "Bennett suggested I get his dissertation on inter-library loan," Kober had written happily to Daniel afterward. "I did! 9 inscriptions, and some very interesting ideas!"

After devouring the dissertation, Kober was confident that the scripts at Knossos and Pylos had been used to write the same language. "There can be no doubt now, I think, that the languages are either identical, or very similar," she wrote to Bennett. What was needed, she knew, was for the two of them to embark on a discreet collaboration, collating their lists of signs and vocabulary into a master list that would cover both Crete and the mainland.

Though relations between the Knossos and Pylos camps had been cool at best, Kober and Bennett were united by similar circumstances: Both had their hands tied by the ego and obstinacy of more senior scholars. As long as Blegen sat on the Pylos inscriptions, Kober could not see them; as long as Myres held on to those from Knossos, Bennett was similarly constrained.

Kober respected Bennett. He was a careful scholar, and had worked as a cryptanalyst during the war, pinpointing patterns in encoded Japanese messages despite the fact that he knew no Japanese. They began to put a tentative modus operandi in place. In a letter of June 7, 1948, Kober formally proposed they work together. "*If* there is a possibility that the publication of the Pylos material will be long delayed, and ditto for the Knossos, it might be worth-while for us to get together on things like sign-list order and content classification. Otherwise, you realize, as soon as both are published, others will start the necessary . . . classifications and all our work will be thrown overboard. If [the publication of the tablets] happens during the coming year, I'll be very happy, but if several years pass (and they have a way of slipping away) it would be silly" to continue working separately.

In the fall of 1948, Blegen at last gave Bennett permission to share the Pylos inscriptions. "Happy, happy day!" Kober wrote on hearing the news. By coincidence, Myres had relaxed the embargo on the Knossos data at around that time. Kober could now share the inscriptions she had copied in Oxford with other scholars privately, he said, though everyone would have to wait until *Scripta Minoa II* came out to publish anything based on them.

"If Bennett is willing," Kober wrote gleefully to Myres at the end of October, "I'll exchange a set of my drawings for one of the Pylos inscriptions. I didn't know I had horse-trading Yankee blood in my veins, but apparently I do."

By the end of the year, after some more delicate diplomacy-by-mail that reads like a negotiation for the exchange of hostages, Kober was traveling to Yale to look at Bennett's inscriptions and he was traveling to New York to look at hers. Bennett, who was then comparing the handwritings of the various scribes and liked to spread out a blizzard of photos for scrutiny, once wrote to ask whether she had "a large, well-tabled room" at her disposal. Kober offered the use of the Ping-Pong table in the basement of her home, but alas, he replied, it would not suffice, for "the last time I did it . . . I eventually covered about three or four ten-foot library tables."

Their first order of business was to compile a definitive sign list—a signary, paleographers call it—containing all the characters used in the mainland and Cretan versions of the script. This was no simple task: The Knossos inscriptions alone still lacked a standard signary after almost half a century. Now, with the Pylos material added, the problem became even more difficult. In the end, the Linear B signary would not be finalized until 1951, a project completed by Bennett and Ventris the year after Kober's death.

"There are a few signs that must be added to our lists, and in my case, at least, all the vocabulary lists must be revised," Kober wrote Bennett that June. "You can imagine what a job it is to go through all the inscriptions and check these things again. . . . I must now convince Sir John Myres that the sign list must be changed to include [the Pylos data], and he does not want to do it. I do not want to do it either, but there is now no doubt in my mind that it must be done."

There were also thousands of vocabulary words to reconcile, and Bennett began compiling a master list of words from both places. He came across many Knossos words that were not found at Pylos—all of which he asked Kober to correct, check, and annotate. These, too, were dispatched to Brooklyn.

* * *

IN 1949, KOBER came out with a noteworthy article that solved two enduring small mysteries in a single stroke. One was the problem of telling male from female animals in paired logograms like 𐂇 and 𐂆. The other was an analogous problem: how to tell apart the two words—𐀒𐀺 and 𐀒𐀷—that almost certainly meant "boy" and "girl" but were likewise indistinguishable. As her paper showed, the same key unlocked both questions, and it had been hiding in plain sight in the tablets all along.

The solution centered on the only Linear B word whose meaning was known beyond doubt—"total." The word appeared again and again at the bottom of inventories on Linear B tablets, and as scholars from Evans onward had noted, the Minoan scribes routinely wrote it in two forms: 𐀵𐀰 and 𐀵𐀭. Though the pronunciation of these forms was unknown, each obviously had two syllables, and the initial syllables were obviously identical. As Kober's major work had shown, it was also clear that the second syllable of each word began with the same consonant but ended with different vowels. But if both forms meant "total," then why, Kober asked herself, did they need to differ at all? She began to look closely at the context in which each form occurred.

As Evans had recognized, some of the inventories on the tablets were lists of names—names of soldiers, names of slaves, names of workers in various trades. Lists marked with the "man" logogram, 𐂀, obviously contained men's names. The famous twenty-four-line tablet from Knossos, known as the "Man" tablet, displays the logogram at the end of each line, just before the numeral:

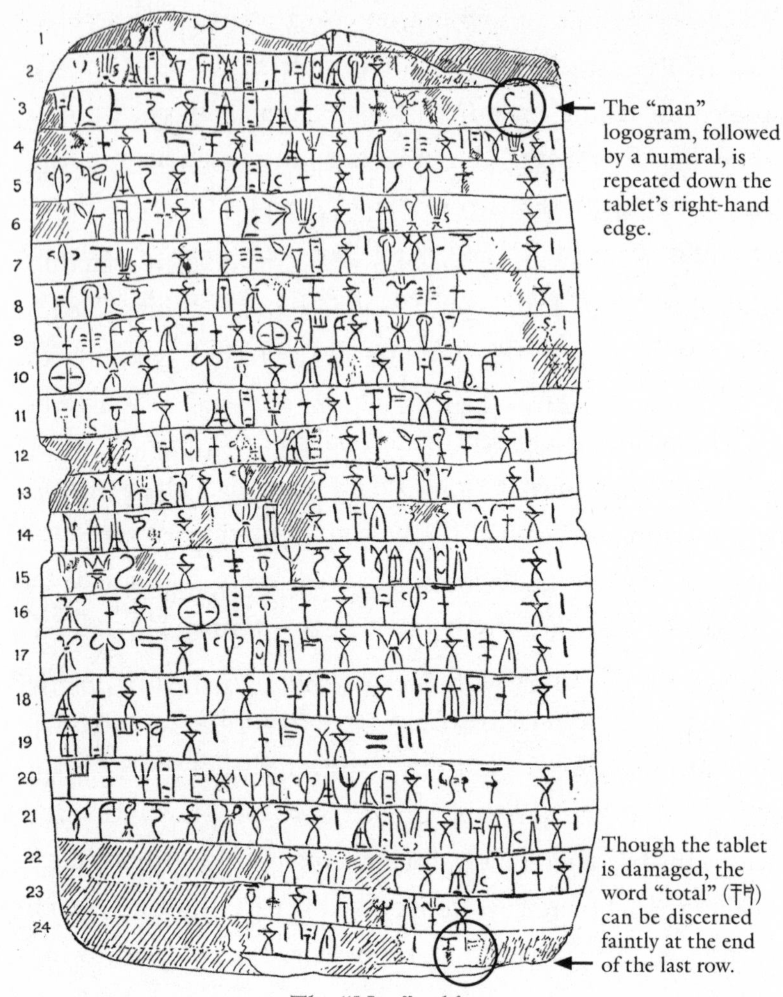

The “Man” tablet.

Other inventories, marked with the logogram 𐂁, listed women’s names.

As Kober studied the inventories, she spied a pattern. Lists of men’s names always used one particular form of “total”: 𐀵𐀰. Lists of women’s names always used the other form: 𐀵𐀭. The difference in form, she realized, could well represent a difference in inflection—specifically, a difference in gender. Many languages inflect their nouns and adjectives for gender: In Spanish, for instance the masculine noun *viejo* means "old man"; its feminine counterpart, *vieja,* means "old woman." Kober deduced that in the language of the tablets, 𐀵𐀰 was a masculine form of "total" and 𐀵𐀭 a feminine.

As she examined the lists of women, she noticed something else. The "woman" logogram was sometimes accompanied by the words 𐀒𐀺 and 𐀒𐀷. As Kober also observed, whenever the word 𐀒𐀺 appeared, the masculine form of "total" was used before the tally. Whenever 𐀒𐀷 appeared, the feminine form was used. Therefore, she concluded, 𐀒𐀺 meant "boy," and 𐀒𐀷 meant girl. (The lists turned out to be records of food rations dispensed by the Knossos palace to slave women and their children.)

By solving the gender problem for people, Kober also solved it for animals. The animal logograms of Linear B also took two forms, one with a barred stem (𐂅, 𐂇, 𐂉, 𐂋, and 𐂍), the other with a V-shaped stem (𐂄, 𐂆, 𐂈, 𐂊, and 𐂌). In lists of animals, she noticed, the masculine form of "total" was used in connection with barred signs, the feminine form with V-shaped ones. With this single observation, she was able to assign meanings to logograms that had remained undifferentiated for nearly half a century:

𐂅 = "stallion"	𐂄 = "mare"
𐂇 = "ram"	𐂆 = "ewe"
𐂉 = "billy goat"	𐂈 = "nanny goat"
𐂋 = "boar"	𐂊 = "sow"
𐂍 = "bull"	𐂌 = "cow"

* * *

AT THE START of 1949, Michael Ventris, amid the press of architectural duties, appeared to withdraw from the fray. "I shall probably give the problem a rest for a bit now, so hurry up and decipher the thing for us!" he wrote Kober that February.

She scarcely had time. Before her second trip to England, Kober had experienced the first glimmer of health trouble. "I just finished my last set of examinations yesterday and my eyes are bothering me—something that never happened before," she had written to Emmett Bennett in June 1948. Around this time, uncharacteristically, she began to make mistakes in her normally careful transcriptions. "About my errors. I must apológize," she wrote in the same letter. "I don't usually make such silly mistakes." To Sundwall, she wrote in February 1949: "I am ashamed at the number of errors I have found in my copy. If there is ever time enough, I will go through it carefully and send you the corrections—but that will not be before this summer."

But by then, she was feeling even worse. In April 1949, when a package arrived containing the initial batch of galley proofs of *Scripta Minoa II,* she set it aside at first. "This year has been a nightmare," Kober wrote to Myres in May. "More and more work at school, and no prospect of a let-up till the middle of June. I haven't had time to do anything else. And, in addition, I am so worn out that for the first time in my life I'm worrying about my health. I hope to recuperate during the summer. One of the reasons I sent the proofs back without correction was that I was feeling too ill to face the thought of checking. I've felt guilty ever since. Here they are."

With characteristic determination, Kober, who inclined toward plumpness, embarked on an uncompromising regimen of dieting in an apparent attempt to restore her health. Meanwhile, she continued to work—on her own research and, increasingly, on that of others. In August 1949, Bennett mailed her a list of Knossos inscriptions about which he had questions. "If you have the time would you go through them and check, say with your blue-green pen all those which you would accept," he wrote her. "The business is probably a full day's work so don't put it ahead of anything important." In retrospect, the request reads like a bitter joke: Kober had less than a year to live.

KOBER'S WORK WITH MYRES—or, more precisely, *for* Myres—was even more time-consuming. For months she had been sending him parcels containing her hand-copied Knossos inscriptions, hemming in the loose pages with things like packaged soup. After each parcel was assembled, there came the chore of mailing it: She lived so far out in the reaches of Brooklyn that the nearest post office was more than a mile away. (Kober appeared, like many lifelong New Yorkers, not to have known how to drive.) To mail a package she either had to walk to the post office or wait until a neighbor had time to take her.

After she finished her hundreds of drawings, Kober's next task was to go over Myres's typescript of *Scripta Minoa II* page by page before it could be dispatched to the Clarendon Press for typesetting. In late 1948 she told Sundwall that she was living with Myres's manuscript night and day, and had to read it very carefully, "because he makes so many little errors." "The truth is," she added, "that he really can't read the Minoan scripts."

In the spring of 1949, the final proofs of *Scripta Minoa II* began arriving in Brooklyn at last—in maddeningly small installments. Whenever new ones showed up, Kober had to drop what she was doing, read them, correct them, and ship them back to Myres. "I never know when proofs are coming," she wrote Bennett in June. "They drive me crazy. I just don't approve of the method. I can correct, but in the end, in my opinion, the whole thing isn't right. Oh, well, it's Sir John's work, not mine. I refuse to take credit or blame for it."

By this time she was utterly exhausted. "School is just finishing," she wrote Myres that month. "I have to have a short rest before I plunge into Minoan again. My health hasn't improved yet—but then, I haven't had a rest yet. . . . I'm a little homesick for England, and for an ocean voyage, but this summer I shall stay here and rest."

Myres also continued to send her batches of the manuscript of *Scripta Minoa III,* the Linear A volume. He was using her, in essence, as a typist, and as Kober's reams of typed correspondence, with their thickets of cross-outs and overstrikes, attest, she was not an especially good one. But for her, the real labor lay not in the typing but in the unrelieved correcting and editing she found his manuscript needed. From this point on in her correspondence, her tone toward Myres grows genuinely bitter.

"Sorry that at times my notes take more space than your text," she wrote him in October 1948, as she returned some pages of *Scripta Minoa III.* "I've just reread them, and am sorry to say that I meant some of them to be funny, but they no longer sound that way." In early November, she told him, "I want to get back to my own job, which is deciphering Linear B. . . . It would be a waste of time at present for me to do this work."

But then, in a postscript, she promised to send Myres "another batch of text soon."

"What I would like to do right now is go to sleep for an entire month," Kober wrote a few days later. Not long afterward, in a letter to Myres on November 28, she tried to abandon the whole project:

> *Now, my reason for this letter. I've stopped working on your manuscript temporarily, because I don't know what to do. . . . It seems to me a waste of both our times (that's a peculiar sentence, but I guess you know what I mean). First I type out your statement, then I check all the references*[,] *then I must write a long note explaining why I think you are wrong, then you must do it all over again, rewrite, and then I'll have to do it all over again. When there are many corrections it takes me a day to do a page of your MS. At that rate it will be months before I finish, and then more time before it's done over. . . .*
>
> *Next week I can't work anyhow, because I'm getting examinations from all my five classes—130 long papers—which will keep me out of mischief for quite a while. . . .*
>
> *I don't suppose that either of us thought it would take so long. I could just type it out without checking anything, but that wouldn't be much help, would it? I'm not doing the checking in my editorial capacity, but because I'm interested both in seeing it published for the sake of scholarship, and in making it as useful as possible.*

Myres asked her to carry on and she agreed, but by the spring of 1949, she had had enough. "I've come to the conclusion that the kind of work I'm doing is useless," Kober wrote him. She continued:

> *It annoys you, for which I don't blame you, though that isn't the effect I'm trying to achieve. But it doesn't result in changing your mind, which I am trying to do. . . .*
>
> *Please let me know what to do with the rest. I can't do with it what I've done up to this time. It would take years. Just reading through it would do no good. . . .*
>
> *I just don't have time for anything except school work. I finished a set of 25 papers to-day, will get a set of 40 on Thursday, another set of 40 on Friday, and two sets on the following Tuesday. . . .*
>
> *My, I am sorry for myself! I'm always in a lovely mood when I finish a set of papers. . . .*

IN EARLY JULY 1949, Kober, feeling especially unwell, went to the doctor. On July 27, she was ordered to the hospital for three weeks' observation. On August 15, she was told she would need surgery, which meant weeks more in the hospital. "I'm sorry," she told Bennett, "but it seems you'll have to mark time till I recover and finish what should have been finished long ago, except for my illness."

Just what illness Kober had is nowhere stated in her correspondence. "I managed to acquire something very unusual, and it took the doctors more than a month to find out what it was—and then they were stumped about a cure," she wrote Myres. It is in keeping with midcentury taboo that a serious illness would never have been named, even in correspondence with valued associates. Nor does any of the published obituaries of Kober list the cause of death, also a customary omission then. Even her death certificate sheds no light on the question: Cause of death is routinely redacted from New York City death certificates of the period. It seems probable, given Kober's heavy smoking, that she had some type of cancer. A much younger cousin, Patricia Graf, who was a child when Kober died, says it was whispered among the women in the family that "Aunt Alice" had a rare form of stomach cancer. (Her father, Franz, had died of the disease.)

It has also been suggested that Kober's doctors never told her precisely how ill she was. Given the low esteem in which the medical establishment of the period held patients—especially female patients—this, too, is possible.

In a short letter from late August 1949, Kober apprised Sundwall of her health, writing in an uncharacteristic huge, childish scrawl: She was too weak, she explained, to sit at a typewriter. "Naturally," she said, "I got no work done on Minoan."

That summer—the summer she was supposed to rest—Bennett continued to send her lists of signs and vocabulary words to check. But over time, Kober began to feel he was getting the better end of the deal. "It is too bad that illness had to strike this summer when I hope to accomplish so much with Minoan," she wrote Myres in late August, when she was housebound after her operation. "My illness will hold up Bennett, but, when I had time to think in the hospital, I realized he is demanding a great deal from me—and so far has given me almost nothing—except I copied *his* copies of the Pylos material. I shall hereafter proceed only on terms of strict reciprocity. He gets my photographs if I get his. Otherwise not."

By autumn, she was worse. In late October 1949, Kober returned from a second, six-week hospital stay and was again housebound. Yet she remained full of plans for Linear B. "Professor Blegen at last relented and I was allowed access to the Pylos material," she wrote Henry Allen Moe of the Guggenheim Foundation on October 29. "I have great hopes, because the new material supplements the old, and there is now almost enough for valid statistical analysis. . . . *Scripta Minoa II* is still held up. I'm beginning to think it will never appear. In another year, I will be ready, if health permits, to start a monograph on Minoan *provided* the material is published. I'll write it in any case, but it may have to wait years before it can be published."

That day, she wrote Bennett: "I hope I am now really on the way to recovery, but it will take some time. . . . As soon as I can, I hope to get back to inflection. . . . Sorry that my illness is holding you up as well, but I can't do a thing about it, except hope." By return mail, Bennett wished her a rapid recovery and asked for her suggestions on the order of the Linear B signary.

By the fall of 1949, Kober was officially on sick leave from Brooklyn College, though she continued her work for Myres. "My health is, unfortunately, not what it should be," she wrote him in November. "I am not bed-ridden, but am at present house-bound, because 10 weeks in hospitals in the last 12 have weakened my legs so that I cannot manage stairs. But enough about that. Naturally, all this has held up my work with Minoan, for which I have high hopes—if I can ever get the preliminary analysis finished."

She dreamed of Crete. "I haven't done anything about going to Greece next summer, since most fellowships, etc., demand a health certificate—which at the moment I could not give," she said in the same letter. "I'll have to have some subsidy now, since I spent all my reserve money on doctors. But I'm sure that if the Museum is open, I'll get there some way."

BY THE END of 1949, Michael Ventris found he could not resist the lure of Linear B after all. Back in the hunt, he launched his largest-scale assault on the script thus far. He drew up a detailed questionnaire and sent it to a dozen scholars around the world, Kober included—and here the pair's stories dovetail for the last brief, painful time.

It had been nearly half a century, Ventris reminded his correspondents, since Evans dug up the first tablets at Knossos. The few serious analysts of the script were working largely in isolation, cut off from one another by threadbare communications technology and the congenital reluctance of scholars to part with trade secrets. "I began to day-dream" about "a symposium from everyone at present working on Minoan language and writing, explaining the position they had reached, and suggesting the next lines of approach—possibly developing into a series of mutual bulletins routed through someone with a duplicator," Ventris had written in 1948. He began to do just that.

What he did, in short, was to create the kind of "mutual aid society" that Kober had envisioned for her Minoan center at Penn: an international clearinghouse through which decipherers could keep abreast of the latest developments.

Like Kober, Ventris saw the decipherment as a collaboration rather than a competition. His approach, as Andrew Robinson points out, owed a great deal to a way of doing architecture, known as Group Working, that was then coming into vogue in Britain. Traditionally, decisions on any architectural project had been made by a single, powerful Great Man—the lead architect—and handed down by fiat to the team's junior members. But by the time Ventris entered the field, architectural offices had begun to adopt a more democratic approach, soliciting opinions from each member of a team, and sharing them among the team as a whole, in a collegial round-robin. (A few years later, Ventris would embark on a project—a prestigious and, as things turned out, searingly painful enterprise—that studied the ways in which information could be disseminated efficiently among architects.)

In preparing and distributing his questionnaire, Ventris appears to have applied the Group Working ethos directly to the problem of decipherment. Among the twenty-one questions he included were these:

- What kind of language is represented in the Linear B inscriptions, and to what other known languages is it related?
- Is the Linear B language identical, or closely related, to that of Linear A?
- What phonetic or other values . . . do you assign to the common signs of the Linear B signary?
- Do you consider that a reasonable degree of decipherment can be achieved?

Of the dozen scholars he solicited, ten replied, including Sundwall, Bennett, and Myres. Only two refused to take part.

The first was the Czech investigator Bedřich Hrozný, who felt he had already deciphered the script. The second was Alice Kober. Her reply to Ventris, dated February 2, 1950, reads, in its entirety: "I have no intention of answering the questionnaire. In my opinion it represents a step in the wrong direction and is a complete waste of time."

Wrenching in its stark brevity, Kober's letter is startling but not surprising. The questionnaire trafficked in the very stuff that galled her, soliciting speculations on the nature of the language and the values of the symbols. And she had truly no time to waste: She was by then incurably ill. (Her position on the questionnaire's usefulness was vindicated, at least in part: Of all the replies Ventris amassed, not one, including his own, identified the language that Linear B actually turned out to write.)

IN JANUARY 1950, with Kober still out on sick leave, Brooklyn College promoted her from assistant professor to associate professor. It was clearly a gesture of charity, for she would not return to school. By February, the references in her letters to her own health, previously optimistic, had grown more measured. Writing to Myres that month, she apologized for returning two sets of *Scripta Minoa II* proofs "at long last."

"It took a very long time to go over them because I can work only a short time each day," she continued. "My doctors are not too encouraging about an early recovery. . . . My work, has, of course, not been progressing. I spend most of my time recuperating."

In a brief, scrawled letter to Sundwall in early March, she

wrote: "I am still sick. I am still not able to leave the house, and find writing letters very hard. Of course, I'll do my best to answer anything dealing with Minoan as soon as possible but please excuse me if I am slow."

That day, in a postcard to Bennett, Kober asked for the return of photographs he had borrowed. The request is painfully ambiguous: It can be read either as a declaration of her continued intent to work with them, or as an acknowledgment that the time had come to put her affairs in order.

On April 4, in another postcard to Bennett, she wrote that she was "still busy checking" his photos and drawings of Pylos inscriptions. "I am still very ill, and may have visitors only 15 min. at a time, so can't ask you to come yet," she continued, adding: "Proofs of SM II arrive sporadically."

That spring, a long article by Kober appeared in the journal *Language.* The article—a joint review of volume 2 of Hrozný's "decipherment" of Minoan together with a comparable "decipherment" by the Bulgarian linguist Vladimir Georgiev—is her last publication. Her impassioned brief for a science of archaeological decipherment, meticulously outlined in the review, reads like an urgent parting shot:

> In the ultimate analysis, a successful decipherment is something that is achieved after many years of hard work by various scholars. It is not, actually, the first step in breaking an unknown language, script, or cipher. *The first step is to find the essential clue.* When a scholar comes upon this, either through fortunate inspiration or by the careful use of scientific method, or more probably by a combination of the

two, others, using the same method, can come to the same conclusions for themselves. As long as a "decipherment" depends on the ingenuity of a single person, whose technique no one else can apply because minds work differently, the essential clue is not available. . . .

For Minoan the clue must be sought in the scripts themselves, and no theory, no matter how attractive, can stand up until it is borne out by incontrovertible evidence from the scripts. . . . If the decipherer starts with the conviction that Minoan is related to Chinese, and ends up with the conclusion that he was right, having "proved" it by translating the documents, he has usually been reasoning in circles. It is one thing to start by considering all the known facts, and to come to a conclusion. It is quite another to start with a preconceived idea, and try to prove it. A scholar's worst enemy is his own mind. Facts are slippery things. Almost anything can be proved with them, if they are correctly selected. . . .

It is unfortunate that it is only in geometry that a scholar must state his assumptions clearly before he begins his proof. . . .

On April 17, 1950, returning the latest set of *Scripta Minoa II* proofs, Kober wrote an astonishing letter to Myres. It is short and to the point, but it is more naked, and more angry, than anything in her entire correspondence. "Dear Sir John," it begins:

Finally, I finished going over this last batch. What a mess! Frankly, if anyone but you had sent it to me, I'd have sent it right back. I never saw so many really inexcusable errors, both in numerals, and, what is infinitely worse, in signs. You were once furious at me for saying you confuse *certain signs—but you do, over and over again. Please get them straight in your mind, because my corrections may not be complete. Also, there are so many errors in the inscription numbers, that a large number of your Plate references may be completely wrong. You'll have to check. I can't.*

Also, I see no trace of the foot-note saying the comments on the classification are yours, not mine. My permission to use the classification is conditional. I will not permit it if you do not make the statement—and make it right there—where nobody can miss it. I disagree violently with some of the things you say and do not wish to spend years telling people I do not believe them. I want neither credit nor discredit for your ideas. . . . You may be right, and I wrong, but I do want things kept straight.

Well, it looks as though SM II should be out pretty soon now. The worst is over.

I am still ill. That is, I am still recovering from the cure. I don't know how much longer it will take. Neither do the doctors.

Then, at the close of the letter, a remarkable about-face:

How are you? And how's the weather? We've been having rather warmish weather here all winter—but had a very cold spell last week, with snow.

My regards to Lady Myres . . . and all the rest of the dear people. I do wish I could get to see you all soon. I really get homesick for St. Hugh's. Isn't it strange? I feel so at home there—more so than at my own college.

Sincerely,
Alice Kober

On the morning of May 16, 1950, Alice Kober died at her home in Brooklyn, at the age of forty-three. The letter to Myres is her last known to anyone. Perhaps that is fitting: For all its pulsating rage, it ends with a vision of Paradise.

BOOK THREE

The Architect

Michael Ventris, 1953. He is copying Linear B inscriptions from Arthur Evans's *Scripta Minoa*.

9

THE HOLLOW BOY

London, 1936

THE KNOSSOS TABLETS, SO LONG in the ground, were under glass now, on display in London at Burlington House, home of the Royal Academy of Arts. Curated by Sir Arthur Evans, the exhibition honored the fiftieth anniversary of the British School at Athens, the archaeological institute of which he was a founder. On this autumn day, Evans himself, at eighty-five the eminence grise of Old World archaeology, was passing through the gallery. There, he came face-to-face with a group of boys on a school trip. They were sixth-formers—sixteen- and seventeen-year-olds—plus one younger schoolmate who had tagged along to see the ancient artifacts.

Evans was fond of children, and he took it upon himself to give the boys an impromptu private tour. They soon found themselves before a glass case that held some of the tablets he had unearthed at Knossos. Despite all efforts, Evans told them, the mysterious writing on them still could not be read.

As was widely recounted afterward, a shy treble arose from

the knot of schoolboys. "Did you say the tablets haven't been deciphered, Sir?" asked the group's youngest member, a boy of fourteen, in polite excitement. The boy's name was Michael Ventris.

Though the episode marked the start of Ventris's obsession with this particular script, he had already been in thrall to ancient writing for half his young life. Endowed with a knife-edge logical mind and an almost unnatural facility for acquiring foreign languages, he was certainly the most natively brilliant of the three major players in the Linear B story. He was also very likely—though here the playing field was practically level—the most obsessed.

But the most remarkable thing about Ventris by far is that he was neither an archaeologist like Evans nor a classicist like Kober nor a scholar in any other field that might have made him a likely candidate to solve the riddle of the script. In fact, he had never even been to college. (In those years, British architectural training took place in professional schools, not in universities, and Ventris's formal education in any subject that might be considered grist for a career in archaeological decipherment had ended in his late teens.)

Like many of the amateurs who tackled the script, Ventris spent years juggling unverifiable speculations, choosing a favorite candidate for the language of the tablets almost at the start. Unlike them, however, his acute rationality, superb gift for pattern recognition, and profound appreciation of mathematically elegant approaches to problem solving of every kind made him, once he finally read and digested Kober's articles, spectacularly well equipped to use her methods to reach a solution.

Since that day at the Royal Academy, Ventris had lived

with the script more intimately than anyone except Kober. At boarding school, he read about it under the covers by flashlight after lights-out. While still in his teens, he wrote fervent letters about it to Evans. At eighteen, he wrote a long analytical article on Linear B that was published in a scholarly journal. In the 1940s, he took it with him to war.

If Linear B exerted a greater hold on Ventris than it did on anyone else, there was ample cause. In his youth, it was a buffer against an unnaturally frigid upbringing. Soon afterward, it became a distraction from great loss: Both his parents had died, one by suicide, by the time he was eighteen. Later, when he was a young architect, it helped leaven his days in an uninspiring job. Ignoring his wife and children and, eventually, his profession, he worked feverishly on the script at every available moment. By the time he was in his mid-thirties, his obsession had cost him his relations with his family, his architectural career, and, in the view of some observers, his life.

THE ONLY CHILD of Edward Francis Vereker Ventris and the former Anna Dorothea Janasz, Michael George Francis Ventris was born in Wheathampstead, a village about twenty miles north of London, on July 12, 1922. (At the time, Kober was in high school and Evans ensconced in his third decade of excavating and restoring the palace at Knossos.) On his father's side, Ventris was descended from an old English family inclined to produce straight-backed military men. His paternal grandfather was a highly regarded army officer who retired in 1920 as commander of the British forces in China. Michael's father, Edward, was an officer in India, though he appears to have had a less lustrous career than his father before him. "Overshadowed by illness and perhaps his own father's military reputation," Andrew Robinson writes in *The Man Who Deciphered Linear B,* "he retired from the Army in his late thirties, as a lieutenant colonel, and was a semi-invalid for most of his remaining years."

Ventris's mother's background was, by all appearances, something of a counterweight. Dorothea, or Dora, as she was known, was the beautiful dark-haired daughter of an English mother and a wealthy Polish landowner who had settled in England. She was passionately interested in the arts, the flourishing Modernist school in particular, and numbered among her friends many of the leading lights of Europe's contemporary art scene. Among them were the designer Marcel Breuer, the architect Walter Gropius, the painter Ben Nicholson, and the sculptors Henry Moore and Naum Gabo.

Edward Ventris suffered from tuberculosis, and Michael spent most of his childhood in Switzerland, where his father had gone for treatment. He attended a boarding school in Gstaad, where he readily acquired French, German, and the local Swiss German dialect. (He had long since acquired Polish from his mother.) Very early on, he became fascinated with the means of writing languages down. At eight, in Switzerland, he bought and devoured *Die Hieroglyphen,* a German-language book on hieroglyphics by the renowned Egyptologist Adolf Erman.

In accounts of Ventris's life, much has been made of his prodigious ability to pick up languages when he was just a boy. In fact, there is nothing remarkable about it: Acquiring languages in childhood is what each of us is biologically predisposed to do. What is genuinely noteworthy is that Ventris continued to do so, with equal facility, to the end of his life.

Children come into the world hardwired to acquire language, provided only that they are exposed to it. From birth through the age of about ten, a child can acquire a first language, as well as a second, third, fourth, and more, by drawing automatically on these inborn principles. Scientists call these years the "critical period" for language acquisition. But then, for neurological reasons that are not well understood, most people automatically exit the critical period near the start of adolescence. After that, new languages are accessible to them only through the arduous classroom instruction with which high school and college students are all too familiar.

But for a charmed few—for neurological reasons that are even less well understood—the critical period seems to continue undiminished through adulthood. They can inhale foreign languages at twenty or thirty as readily as they did at six, with minimal effort. Michael Ventris was beyond doubt such a person, possessing a rare, innate gift that would serve him singularly well in the years that followed.

"How do you come to be so expert in Swedish?" his collaborator John Chadwick wrote him in the 1950s, after Ventris sent him his own translations of accounts of the decipherment in the Swedish press. (The answer was simple: Ventris had mastered the language as an adult, in a matter of weeks, in preparation for a short-term architectural project in Sweden.)

In 1931, when Michael was about nine, his family returned to England, although the coming years would be punctuated by stays on the Continent for Edward's health. In England, Michael was sent to Bickley Hall, a preparatory boarding school in Kent. On school holidays, he came home to a house of little warmth. As Robinson and others have remarked, the Ventrises displayed a coldness toward their only child that was remarkable even for their time, place, and class. There was a reason: In Switzerland, Edward and Dorothea had undergone psychoanalysis with Freud's disciple Carl Jung, and their treatment of their son was per Jung's instructions, intended to prevent Michael from forming an Oedipal attachment.

As Jean Overton Fuller, the daughter of a Ventris family friend, recounted in *A Very English Genius,* the 2002 BBC documentary about the decipherment: "Colonel Ventris said we must come and meet Michael. But Mother gave me a warning: I must on no account touch Michael. It had been explained to her that Michael was never to be touched by anybody. This was to avoid his having complexes." She added: "What my mother was afraid of was that he would never be able to make a natural, warm relationship, never having had one."

Fuller, who was nine at the time of the visit (Michael was two), later wrote of overhearing Edward Ventris confide in her mother, whom he had known in India:

> I heard it all as Colonel Ventris poured it out to my mother, unembarrassed by my silent presence. Probably I was deemed too young to understand. But what I heard him say to my mother was that Jung said Dora's father, a Polish count, was a tyrannical autocrat who bullied Dora, ordering her about, terrifying her . . . but that subconsciously Dora was in love with him and hoped to find him again in her husband, but he, poor husband, just hadn't got it in him so to mistreat her. He couldn't stop being patient and considerate and it irritated her, she felt he wasn't her idea of a man.

Jung, Fuller wrote, "had complicated matters by falling in love with Dora, so that . . . it was improper he should entertain such feelings for her and continue to treat them both as his patients." He discharged both Dorothea and Edward from his practice. "So then they were left stranded," Fuller wrote. "My personal feeling is that whatever they were like before they met Jung, he made them not better but much worse, over engrossed in the analysis of their complexes, depressed."

In 1935, at thirteen, Michael was sent to the Stowe School in Buckinghamshire, about fifty miles northwest of London. A progressive secondary school, it emphasized culture and the classics over athletics, a time-honored staple of English boarding school education. In the BBC documentary, a classmate, Robin Richardson, recalled the young Ventris as "an aloof, slightly abused and detached member of our own community who didn't make great friends with the rest of us. . . . He regarded all of us with a degree of puzzlement and amusement." Ventris himself has written of his school days, "I think they rather thought me a black sheep while I was there and it'd be rather insincere to make out I had any deep tradition implanted in me!"

His teachers recalled him as a middling student who expended little effort on subjects that did not interest him. But it was at Stowe that Michael joined the fateful class trip to the Royal Academy in London. It was at Stowe, too, that night after night, after lights-out in the dormitory, he pored over the few available transcriptions of Linear B tablets, most likely from volume 4 of Evans's *The Palace of Minos,* published in 1935.

By this time, Edward and Dorothea's marriage, a strange match from the start, had taken an even stranger turn. From about 1932 on, as Jean Fuller recalled, the couple lived in Eastbourne, on England's south coast, occupying separate hotels along the seafront. They would meet each day on a bench in between the two. "If they were apart, they pined for each other's society, but if they were together, they hurt each other," she said. "They were like porcupines." The couple formally divorced when Michael was about fourteen. Edward died two years later, in 1938.

After the divorce, Dorothea and Michael moved into a newly erected London apartment building known as Highpoint. A blocky tower of white concrete, Highpoint had been designed by the Russian émigré architect Berthold Lubetkin to be the apotheosis of Modernism. Inside, their flat gleamed with work by the modern artists Dorothea cherished, many of them her personal friends. There were two Picassos on the walls, as well as sculptures by Gabo and Moore. From Breuer, she commissioned much of the furniture, with its clean lines, light wood, glass, and chrome. The collection he designed for her included a small glass-topped sycamore-and-chrome desk, now in the permanent collection of the Victoria and Albert Museum, on which her son would one day decipher Linear B.

Dorothea and Michael loved the clean, modern aesthetic of their new home. "I count myself extraordinarily fortunate to have this little centre which you made . . . your interior with its shapes and colours and textures of which I never tire," Dorothea wrote to Breuer. "Other people appreciate it too, but no one as much as Michael, who has such a firm affection for his room that I am sure he will never let me give up the flat."

All the while, Michael continued his investigations into Linear B. He had already taken it upon himself to write to Evans with his theories on script, a correspondence he initiated not long after his life-changing class trip.

"Dear Sir," begins one letter, of twenty-three pages, written in the spring of 1940, when Ventris was not quite eighteen: "I don't know whether you remember my writing to you a few years ago about some theories I had on the elucidation of Minoan. Actually, I was only fifteen at the time, and I'm afraid my theories were nonsense; but you were very kind and answered my letters. I was convinced that the key would prove to be in Sumerian, but am glad to say I have given these ideas up long ago. However, I have continued to work at the problem off and on, and I am coming round more and more to the view that the language contained in the inscriptions is a dialect closely related to Etruscan."

Ventris went on to flesh out his new theory in minute philological detail. He also posited sound-values for many Linear B characters, and meanings for entire words and phrases. In closing, he wrote:

> *I have been interested in the Minoan inscriptions for exactly four years now, not very long perhaps but long enough to make me tremendously intrigued and impatient to see what the eventual outcome will be, and, whatever the approach that may best prove to be the right one, I am convinced that now, more than ever before, is the time for a decisive and concerted effort to liquidate the problem.*
>
> *Yours sincerely,*
> *Michael Ventris*

Around this time, Ventris began work on a scholarly article about Minoan, which he would mail to the *American Journal of Archaeology.* (He would discreetly neglect to tell the editors how old he was.) He and Dorothea were also planning his post-secondary schooling. Though it is seductive to imagine Ventris reading classics at Oxford or Cambridge, that scenario would never come to pass: There was no money for a university education. Since her divorce, Dorothea had depended crucially on income from her family's holdings in Poland. After Germany invaded Poland on September 1, 1939, the Janasz lands were seized. Lacking tuition, she had already been forced to withdraw Michael from Stowe before he could graduate. University, too, was now out of the question.

But Dorothea's love of aesthetics, which she had passed on to her son, would provide an alternative course: Michael could train for a career in architecture, bypassing a university stay altogether. He wrote to his mother's friend Marcel Breuer for advice on where to apply. Breuer suggested the Architectural Association School of Architecture, known as the AA, then as now an independent professional training ground in London. Ventris enrolled there in January 1940, at seventeen.

As the war engulfed England, Dorothea, fragile at the best of times, grew increasingly depressed. "She had already lost her brother in the First World War and her husband was dead," Andrew Robinson has written. "Now her father was a refugee in London and her only son . . . faced the prospect of military call-up." On June 16, 1940, less than a month before Michael's eighteenth birthday, Dorothea took an overdose of barbiturates while staying at a seaside hotel in Wales. "The coroner's verdict," Robinson reports, "was 'suicide while the balance of her mind was disturbed.' " To the end of his life, Robinson added, Ventris would never speak of what happened. After his mother's death, he stayed for a time with family friends, then returned to Highpoint alone.

IN THE MONTHS that followed, Ventris continued his studies at the AA. He also worked feverishly on his Minoan article, doubtless a welcome distraction. He tore up two drafts before he was satisfied. In December 1940, the eighteen-year-old Ventris sent it off to the *American Journal of Archaeology* with the following cover letter:

> *Dear Sir:*
>
> *I am enclosing an article which I should be pleased if you would consider for publication in your Journal. It contains the results of five years' research on the language and writing of the Minoan civilisation, and is intended as a prelude to a full decipherment of the inscriptions.*
>
> *I have elaborated and tried to confirm the theory that the pre-hellenic language of the Aegean is a dialect closely related to Etruscan, and I am confident that along these lines a full solution of this outstanding problem will be possible.*
>
> *I had intended to make the article somewhat shorter, but, when finished, I found I was unable to cut it down below its present length, of about 15.000 words, without leaving out essential material. . . .*

The journal accepted the paper at once, publishing it in its last issue of 1940 as "Introducing the Minoan Language," by

M. G. F. Ventris. Rife with speculation, it was, as the classicist Thomas Palaima points out, "close to worthless . . . both in its unsound methods and in what Ventris' own decipherment would demonstrate was its erroneous guesswork." Except for positing an Etruscan-like tongue as the language of the tablets—a reasonable surmise given geographic and historical realities—Ventris's article is not all that different from the writings of some of Kober's more outlandish amateur correspondents.

It is striking, Palaima notes, that in the large bibliography of articles about Linear B that accompanies Kober's 1948 paper, "The Minoan Scripts: Fact and Theory," she chose to omit Ventris's 1940 article entirely. Perhaps she did so out of contempt for the paper, so clearly the work of an amateur. Or perhaps, as Palaima suggests, she did so as an act of mercy, to spare the adult Ventris a reminder of his juvenile effort.

WHILE STUDYING AT the AA, Ventris became romantically involved with a classmate a few years older than he, Lois Elizabeth Knox-Niven, known to her friends as Betty, or Betts. A letter he wrote to a family friend, the sculptor Naum Gabo, in early 1942, when he was nineteen and his personal circumstances were about to change dramatically, displays his characteristic mixture of reserve, candor, and dry, detached wit:

> *It looks as if, in the ordinary way, we'll have a baby some time round next November—at least Betts has changed her mind, and she wants to have it, and I don't think either that it would be quite right to stop new life when this world needs it*

> *so, quite apart from the risk. But the social politics of it is all rather involved, and we're in the process of working that out. So we* might *get clandestinely married—but all this in confidential in the extreme!*

Ventris and Lois married in London in April 1942; their son, Nikki ("the nicest present St. Nicholas ever brought," Ventris wrote), was born in early December.

By the time his son was born, Ventris was at war: Called up in the summer of 1942, he had joined the Royal Air Force. His letters home to Lois from his training camps, first in England and later in Canada, betray his formal, almost Holmesian, rationalist mien; insatiable curiosity; ardent youthful romanticism; impassioned liberal humanism; fundamental discomfort around other people; and, beneath all this, a deep wellspring of tenderness. In a letter from his camp in Yorkshire composed shortly after Nikki's birth, the nineteen-year-old Ventris writes:

> *Darling Lois,*
>
> *. . . My latest job is making lots of little labels with numbers on for a board in the orderly room on which they indicate who's here, in hospital, on leave, etc.* [Here Ventris has sketched a picture of one such label: A16. It appears to have been a forerunner of the tags he would later make to help him sort the symbols of Linear B.] . . .
>
> *In between, & in the evenings, I've been doing a lot of self-education, but find myself getting more and more irritated by little mannerisms of other people. They bring up food into the reading room, & their prolonged heavy breathing, clopping & wind annoys me, & there are others*

who will jog the table all the time & leave the doors open and tap out rhythms with their feet. . . .

But all the same I make for myself as far as I can a mental oasis, because I feel in the mood for some thinking just now. I think Nikki is the chief cause. I've purposely left my convictions vague during the last few years—but being a father makes one want to make one's ideas more definite. I can only do that by gradually finding what facts & concepts fall into a progressive pattern for me, and to do this I feel I have to skim through lots of different books on lots of different subjects, and I feel I'm wasting time if I read fiction and things which don't mirror life documentarily. . . . One feels one owes it to one's children to at least have full knowledge oneself, and so to be able to start them right. That's why I want to know the essentials about all these subjects:

What matter is ultimately made of, and elementary physics & chemistry; the development of the earth; how plants & animals work, and how they have developed from the old single cell; how man's developed, by evolution, and by being able to make tools and organise himself socially; the present outline of the world, what different countries look like, and produce; the way different countries live, the way our civilisation developed & is now; the existing mechanism of our tools & inventions, and how they're made & work—and the same with all our man-made things; the way our bodies function; the way our minds work; and finally, the organic conditions for future evolution.

It's a big program, and not all of it immediately puttable across to a child. But I feel that a responsible person who doesn't have a clear general picture of all this,

> *the outside world as it was & is, [together with several basic languages, to be able to understand other people] is fundamentally ignorant, and can't help being biassed more than necessarily in his outlook to the world. . . .*
>
> *That's my attitude anyway, just now. I'm sorry it sounds like preaching, but a letter's inevitably a rather one-sided discussion. And, as a mental picture, you and Nikki are sitting in a very nice glow, right in the foreground, where I love you most.*

Ventris's "self-education" program also included learning as many new languages as he could. One was Russian, which he was teaching himself in order to write to Gabo in his native tongue. "My knowledge is gradually getting on," Ventris reported to Lois, "though I can only give a couple of hours a week to it. Still, soon I'll be able to read most simple stuff straight off. When I know Spanish as well [& that's the easiest to learn of them all] that'll be the 5 European ones spoken by the most people in the world."

Nor was he neglecting Linear B. Ventris was serving in the RAF as a navigator, and the job suited him: All about maps and mathematics, geometry, logic and reason, it interested him far more than actual flying did. "It's a desk job, really, in the middle of the plane," an AA classmate, Oliver Cox, recalled his having said. Returning to England after his Canadian training, Ventris took part in bombing runs over Germany. Navigation came so easily to him that, as the British journalist Leonard Cottrell has written, "on one occasion he horrified his captain by navigating his way back from Germany with maps he had drawn himself. On other raids he would set course and then, clearing a space on the navigator's table, happily set to work on his Linear B documents, while the aircraft groaned its way home, searchlights stretched up their probing fingers, and bursts of flak shook the bomber."

At war's end, Ventris hoped to meet with Myres in Oxford and see Evans's transcriptions firsthand. But because of his foreign-language prowess, he was kept on for another year to help interrogate German prisoners. Finally discharged in the summer of 1946, he returned home to his wife and children (a daughter, Tessa, had been born that spring) and resumed his studies at the Architectural Association. He also met with Myres and began copying inscriptions for publication.

IN THE SUMMER of 1948, Ventris and Lois graduated with honors from the AA and were now qualified architects. To celebrate, they parked the children with her family and, with their classmate Oliver Cox, set out on a grand tour of the European continent. But when they reached the south of France, Ventris abruptly insisted on returning to England, much to his companions' surprise. A letter from Myres had caught up with him, summoning him to Oxford for six weeks' intensive work preparing *Scripta Minoa II* for the printer. Ventris went there at once, joining Myres and Kober, then in the middle of her second visit.

But after just a day or two, Ventris fled. His real reasons for backing out of the project were known only to him, but it seems fair to assume that alongside the eminent archaeologist Myres and the brilliant philologist Kober, Ventris was painfully conscious of his status as an amateur.

What is known is that soon after arriving in Oxford, Ventris returned to London, mailing a brutally self-lacerating letter to Myres from the train station on his way out of town. It anticipates a letter he would write eight years later, in 1956, shortly before his death—again involving a retreat from an important project, and again shot through with doubt, pain, and shame.

Ventris's 1948 letter, dated only "Monday night," begins, "Dear Sir John":

> *You will probably think me quite mad if I try & account for the reasons why I'll be absent on Tuesday morning, & why I should like to ask either Miss Kober, or the other girl that you mentioned, to complete the transcription.*
>
> *One would have thought that years in the forces would have cured one of irrational & irresistible impulses of dread or homesickness. But however much I tell myself that I am a swine to let you down after all my glib promises & conceited preparations,—I am hit at last by the overwhelming realization that I shall not be able to stand 6 weeks work alone in Oxford, & that I am an idiot not to stick to my own last. Perhaps it's greater weakmindedness to throw in the sponge, than to grind on with something one's liable to make a botch job of—I don't know. In any case, I shall await Scripta Minoa with great interest—and be too ashamed to look inside. . . .*
>
> *I hope this letter will arrive soon enough to relieve you of any unnecessary anxiety, & that in time my precipitate retreat be not too harshly judged.*
>
> *Yours sincerely,*
> *Michael Ventris*

Myres's reply has not been preserved, but it is clear that Ventris was forgiven: At Myres's request, he would continue copying batches of Linear B inscriptions from his home in London for several years to come.

As Myres made plain throughout his correspondence, he was immensely pleased with Ventris's copying. But his regard for his young colleague's abilities evidently went deeper still—despite Ventris's own recurrent self-doubt. In 1950, the journalist Leonard Cottrell, who wrote often about archaeology, visited Myres at his home. The talk turned to the script, still undeciphered after half a century.

"The man who may decipher Linear B," Myres told Cottrell soberly, "is a young architect named Ventris."

10

A LEAP OF FAITH

IN SEPTEMBER 1949, MICHAEL VENTRIS took a job with the British Ministry of Education in London. He had been hired to help design new schools, part of the collective effort to rebuild the nation after the war's devastation. As those who knew him attest, he was as brilliant in his vocation as he was in his avocation. In her memoir *The Morning Light*, the journalist Prue Smith recalled meeting Ventris during this period. "He was," she wrote, "a very gifted architect with a particular power of analysing the complex sets of data which school building requires and the many constraints which in the difficult post-war days it had to observe—all unseen things which architects have to attend to before the design process can begin."

Smith could just as well have been describing the sifting and sorting and analysis that underpins a successful archaeological decipherment. Besides his remarkable interpretive powers, Ventris also brought to his profession an extraordinary capacity for innovation. With his colleague Oliver Cox, he invented what Smith describes as "a strange architectural drawing aid" that was "made of transparent plastic and resembled a young giant sting-ray, with a broad, rounded head and a long tail." The device, she wrote, allowed an architect "with instant accuracy, instead of sheaves of calculations, to draw perspectives of the insides or outsides of buildings."

At the same time, Ventris conducted his on-again, off-again affair with Linear B. Though he had appeared to bow out in February 1949, exhorting Kober to "hurry up and decipher the thing," by autumn he was working feverishly on the script during lunch breaks at the ministry, much as he had worked in his "office" at the back of the plane on wartime bombing raids. Before long, he found his day job could not compete with the lure of the script. "It is hard to see how designing schools could have excited or challenged his mind for long," Andrew Robinson wrote, "especially as he . . . had not much enjoyed his own school days, and was not very interested in children."

By the end of 1949, Ventris had received ten answers to the questionnaire he had circulated and was working feverishly to collate them. He combined them into a single large document, translating the replies as needed from French, German, Italian, and Swedish. The result was a thick state-of-the-field summary he titled "The Languages of the Minoan and Mycenaean Civilizations, New Year 1950," which came to be known informally as the Mid-Century Report. At his own considerable expense, he duplicated and mailed it to each scholar.

The last section of the report contains Ventris's own detailed answers to his twenty-one questions. But at the end of his contribution, he bowed out of the hunt yet again, writing: "I have good hopes that a sufficient number of people working on these lines will before long enable a satisfactory solution to be found. To them I offer my best wishes, being forced by pressure of other work to make this my last small contribution to the problem."

Ventris kept his word at first, pursuing little but architecture for the next several months. Then, in the summer of 1950, he had the opportunity to meet Emmett Bennett in London: Bennett had come to England from Yale to help Myres prepare the *Scripta Minoa II* inscriptions for publication, a task he assumed after Kober's death. Ventris and Bennett, who shared a similar dry wit, developed an immediate rapport. The meeting also gave Ventris the chance to see some of the unpublished inscriptions from Pylos.

After that, Ventris was irretrievably in the grip of the script. He had made out handsomely in the stock market (in the course of a single transaction, Robinson reports, he earned more than he did in a full year at the ministry) and soon quit his job to devote himself entirely to decipherment. Now his real detective work begins, and within eighteen months—between the start of 1951 and the middle of 1952—he would solve the riddle of the script, a solution arrived at through a combination of genius, perseverance, and remarkably inspired guesswork.

DURING THIS PERIOD, Ventris chronicled his progress, with its attendant false starts, hunches, and minute advances, in a series of typed documents he called Work Notes. There would be twenty Work Notes in all, together comprising nearly two hundred pages; these, too, he duplicated and sent to his circle of scholars. The Notes are profuse and often hard to penetrate, filled with statistics, linguistics, etymology, geography, and Ventris's speculations on everything from the formation of words in the language of

Linear B to the movements of populations in the ancient world. In the first Note, dated January 28, 1951, he flails more or less blindly. In the last, dated June 1, 1952, he solves the riddle.

Work Note 1 does have two notable features. First, Ventris independently replicates Kober's "slight discovery," from late 1947, about the sign 𐀤, which functioned as "and" when attached to the end of a noun. Second, the note contains his own first attempt on paper at a phonetic grid for the script. (Ventris had previously built a three-dimensional "grid," now lost, consisting of a board studded at regular intervals with hooks or nails; from these he hung paper tags bearing Linear B symbols. The tags, which almost certainly recalled those he had made for the orderlies' room in the RAF, could be repositioned at will, with symbols assigned to particular rows or columns according to the consonants or vowels they were believed to share.)

In Work Note 1, Ventris sets down his first written grid, modeled on the one in Kober's famous 1948 article. Her grid, repeated here, was small in scale. It was also refreshingly abstract, plotting only *relative* values among the signs while refusing to assign hard-and-fast sound-values to any of them:

	Vowel 1	**Vowel 2**
Consonant		
1	[Linear B sign]	[Linear B sign]
2	[Linear B sign]	[Linear B sign]
3	[Linear B sign]	[Linear B sign]
4	[Linear B sign]	[Linear B sign]
5	[Linear B sign]	[Linear B sign]

Ventris's grid was far more ambitious. Where Kober had confined herself to ten signs, he, in his crisp architectural hand, plotted twenty-nine. He also rearranged many of Kober's original ten, which was a mistake: As the decipherment would show, she had plotted the relationships among those signs with 100 percent accuracy. Worst of all—and this would have had Kober spinning in her grave—Ventris proposed sound-values for all twenty-nine of his signs. They seemed to derive largely from analogies to sounds in ancient Etruscan, for he was still committed to an "Etruscan solution" to the problem. It was the position he had put forth in his youthful article of 1940, and one to which he would hold fast until only weeks before his decipherment. Ventris's first grid "must be regarded as a failure," Maurice Pope writes in *The Story of Archaeological Decipherment.* Nearly 70 percent of its values were wrong.

Then, in the spring of 1951, Ventris received a copy of Bennett's newly published book, *The Pylos Tablets.* Long awaited, it represented the first substantial published transcription of Linear B tablets anywhere; *Scripta Minoa II,* containing Evans's Knossos tablets, remained unpublished. (When Ventris went to the post office to pick up the precious package, as Robinson recounts, "a suspicious . . . official asked him: 'I see the contents are listed as PYLOS TABLETS. Now, just what ailments are pylos tablets supposed to alleviate?' ")

The Pylos Tablets contained the first established signary for the script, an inventory of its more than eighty syllabic symbols, carefully teased out from the tangle of signs drawn on clay. Having a signary made serious analysis of the script possible at last, for without it, investigators were little better off than our

'B' SYLLABARY PHONETIC 'GRID'

Fig. 1
MGFV

1: State as at 28 Jan 51 : before publication of Pylos inscriptions.

CONSONANTS	Vowel 1	Vowel 2	Other vowels ?	Doubtful
	NIL ? (-o ?) = typical 'nominative' of nouns which change their last theme syllable in oblique cases	-i ? = typical changed last syllable before -ξ and -日.	-a, -e, -u ? = changes in last syllable caused by other endings. (5 vowels in all, rather than 4 ?)	
1 t- ?	ag	aj		ax (Sundwall)
2 r- ??	az	iw	ah ol	
3 ś- ??	eg	aw	oc oj	
4 n- ?? s- ??	od	ok	ib	is oh
5		ak ⇄?	ef	
6 l- ?	ac	ij		
7 h- ??	ix		if	
8 θ- ??	en		id	ex
9 m- ? k- ?	ay –	if an enclitic	"and".	al
10				om av
11				
12				
13				
14				
15				
		aj ij ak il aw og ej oh er oj ex ok ib iw	← group of syllables, including those occurring before -日 on 'woman' tablet (Hr 44, PM fig 689), and those characteristic of alternating endings -ξ & -日. About ¾ of these 14 signs very likely include vowel 2.	

Ventris's first grid. His proposed sound-values for consonants run down the leftmost column. (The two-letter designations, like "ag," "az," and "eg," beside each character should be ignored: They are not sound-values but rather a shorthand key Ventris used to classify the symbols.)

bewildered alien, adrift in Times Square. The Linear B signary has long been attributed to Bennett, and with ample justification, for he worked for years to compile a definitive list. But it is also clear from his long correspondence with Kober on the subject—a blizzard of back-and-forth about which signs were the same and which authentically differed—that she deserves as much credit as he.

Bennett chose to arrange his signary by character shape, with simpler, straighter characters coming first and more complex, curvier ones coming later; until true sound-values could be assigned it was as good an organizing principle as any. His signary looked much like this:

The syllabic signs of Linear B, with characters of similar shape grouped together.

Note that some symbols, like the little pig in the last row, function both as logograms and as syllabic characters in Linear B. Other writing systems likewise conscript symbols for double duty: In the Roman alphabet, *K* can be phonetic, standing for the hard "c" sound, or (as used by a baseball-loving American) it can be logographic, denoting the word *strikeout.*

Now, with a bounty of inscriptions to work from, Ventris could start to make real progress. He began to count the Pylos characters, gradually compiling a set of statistics on their frequency. This let him impose order on the unruly mass of inscriptions, and this newly imposed order led him to his first truly significant discovery.

Ventris sorted the Pylos characters into three groups: Frequent, Average, and Infrequent. Much as Kober had done with the Knossos inscriptions, he also tabulated their use in various positions in Linear B words. As a result, he saw something striking: Five characters—𐀀, 𐀁, 𐀂, 𐀃, and 𐀄—appeared especially often at the beginnings of words.

There are different types of syllabaries, with the type depending on how large a syllable each character represents. The Linear B syllabary, as investigators had long known, was a "CV" syllabary, with each character standing for one consonant plus one vowel. A CV syllabary can spell most words adequately—as long as they start with a consonant. But what happens when a word starts with a vowel? Ventris realized that the five characters he had isolated were the exceptions to the one-consonant-plus-one-vowel rule: They were used to write the "pure vowels"—"a," "e," "i," "o," and "u"—found often at the beginnings of words.

He would eventually identify these characters as follows.

(Note the presence of 𐀃, whose "throne-and-scepter" shape had seduced Evans into calling it a determinative):

𐀀 = "a"
𐀁 = "e"
𐀂 = "i"
𐀃 = "o"
𐀄 = "u"

The discovery was Ventris's first meaningful advance in assigning sound-values to Linear B characters. It gave him grist for his grids, of which he would eventually produce two more. One, in Work Note 15, from September 1951, plotted fifty-one signs. (In this grid, Ventris prudently refrained from assigning sound-values.) The other, in Work Note 17, from February 1952, also plotted fifty-one signs and again posited sound-values. These values, Ventris wrote, "represent the values which seem the most useful in giving an 'Estrusciod' character to Pylos names, words and inflexions."

Histories of the decipherment have focused extensively on Ventris's grids, but what is truly significant is that they helped speed the decipherment *despite the fact that they were only partly correct.* Even in his third grid, Pope observes, Ventris has assigned wrong values to 25 percent of the signs. But he was nonetheless able to crack the code—an even more extraordinary achievement in light of the imperfect road maps he had constructed.

How, then, did Ventris manage to do it? The answer lies in his making a single intuitive leap, which would provide the catalyst for the whole decipherment. It was a deeper version of the kind of leap needed to equate the symbol [symbol] with the English word *merry* in the Blissymbolics problem we examined earlier.

Ventris's leap centered on the sets of related nouns he called "Kober's triplets." When he first read her articles in 1948, he wasn't persuaded that the triplets truly made the case for inflection. "I must say I was slightly disappointed," he wrote Myres then. "I envy the orderly presentation, but I don't feel it results in conclusive proof that the 'terminations' are actually inflexional." Ventris thought that the triplets more likely represented "alternative name-endings," on the order of *Brooklyn/Brooklynite/Brooklynese,* and so forth.

Little by little, Ventris came around to Kober's point of view, and before long he was combing the Pylos data for additional triplets and "near-pairs"—words or phrases, including [Linear B signs] and [Linear B signs], that differed in only one or two final signs. By the summer of 1951, he wrote, he had collected 160 such groups, "which on the face of it show inflexion." These related forms helped Ventris isolate still more "bridging" characters of the kind Kober had described, and to add the corresponding sound-relationships to his grid. Ventris at last seemed to realize that Kober had handed him the key—the critical intersection of inflected language with syllabic script, and the "sharing" of consonants and vowels that results. Now he had only to figure out which door the key unlocked.

Before he did so, he made time for a trip to Crete and the Greek mainland in the fall of 1951. In Istanbul for a conference on the ancient Near East, he traveled on to Crete, where he stayed at the Villa Ariadne and saw Knossos. "I was frankly rather disappointed in the Palace & in the frescoes in the Museum which seemed to me to have been restored in such a way as to kill most of their charm & atmosphere," he wrote to Myres. "Mycenae, on the other hand, where there's little else but rubble, the atmosphere is stupendous."

Returning to London, Ventris began to take hesitant steps toward his second major advance. Now that he'd been through the Pylos data, he realized something vital: While it contained many "triplets" of the *kind* Kober had identified, the *specific* triplets she had isolated were found only at Knossos. The words unique to Knossos included a set she had published in her 1946 paper on declension:

	(a)	(b)	(c)
Case I:	[Linear B]	[Linear B]	[Linear B]
Case II:	[Linear B]	[Linear B]	—
Case III:	[Linear B]	[Linear B]	[Linear B]

What sort of words, Ventris wondered, might be particular to one place but not another? For the answer, one has only to look to any modern municipal document. The municipal documents in my life come from Manhattan. The diverse official papers I've accumulated here over the years, from marriage certificate to local tax forms to a string of jury summonses, all have one thing in common: Somewhere on each of them, prominently featured, appear the words "City of New York." The Cretan tablets were the administrative documents of the palace at Knossos. Perhaps, Ventris conjectured in early 1952, the words found exclusively there were the names of Cretan towns. With that in mind, he tried the following experiment:

Ventris homed in on the simplest forms in Kober's paradigm, the three words—𐀒𐀜𐀰, 𐀶𐀪𐀰, and 𐀀𐀖𐀛𐀰—in the bottom row. Examining 𐀀𐀖𐀛𐀰, he already assumed the first character was the pure vowel "a." By making "only a little adjustment" to his grid, as he told Myres, he was able to plug in reasonable guesses for the values of other signs, much as Champollion had done in the famous "Ramses" cartouche. What Ventris wound up with looked an awful lot like the Greek names, spelled syllabically, for three major towns of Cretan antiquity: Amnisos, Tulissos, and Knossos.

That didn't mean, of course, that the language of the tablets was Greek: Though all three names were attested in Classical Greek writings, they were known to be of pre-Greek origin. Nevertheless, even the barest hint of Greek raised a distasteful enough specter for Ventris to reject his results out of hand. He still believed firmly in an "Etruscan solution" and remained adamant, as he'd written as an eighteen-year-old in 1940, that "the theory that Minoan could be Greek is based . . . on a deliberate disregard for historical plausibility." He dismissed the place-names as a mirage and cast the idea aside.

11

"I KNOW IT, I *KNOW* IT"

IN THE SPRING OF 1952, *Scripta Minoa II* was published at last. (Though Myres had written Kober in 1948 that "all your help in the B volume will of course be acknowledged in the preface to it, very gratefully," the published book gave scant indication of the full extent of her years of hard labor.) The volume, with its hundreds of inscriptions, gave Ventris still more data. In May, he tried his place-name experiment a second time, with one significant difference: He allowed himself to make use of a clue he had previously shunned—the only external clue in existence to the possible identities of some Linear B characters.

The clue had been available to investigators from the very beginning, but as they all knew, it was a risky one. It came in the form of an Iron Age writing system known as the Cypriot script. Also a syllabary, the script was used to record the indigenous language of Cyprus between about the seventh and the second centuries B.C. (The Cypriot syllabary was a descendant of the Cypro-Minoan, on which John Franklin Daniel had done important research in the early 1940s.) The *language* of the Cypriot script remained a mystery, but the sound-value of each character was known: After the Hellenization of Cyprus, the syllabary was retained for a time to write Greek, and the island teemed with Greek syllabic inscriptions on monuments and coins. Thanks to the discovery in 1869 of a bilingual inscription in Cypriot and Phoenician, the Cypriot syllabary was deciphered in the 1870s.

𐠀 = "a"	𐠁 = "e"	𐠂 = "i"	𐠃 = "o"	𐠄 = "u"
𐠊 = "ka"	𐠋 = "ke"	𐠌 = "ki"	𐠍 = "ko"	𐠎 = "ku"
𐠭 = "ta"	𐠮 = "te"	𐠯 = "ti"	𐠰 = "to"	𐠱 = "tu"
𐠞 = "pa"	𐠟 = "pe"	𐠠 = "pi"	𐠡 = "po"	𐠢 = "pu"
𐠏 = "la"	𐠐 = "le"	𐠑 = "li"	𐠒 = "lo"	𐠓 = "lu"
𐠔 = "ma"	𐠕 = "me"	𐠖 = "mi"	𐠗 = "mo"	𐠘 = "mu"
𐠙 = "na"	𐠚 = "ne"	𐠛 = "ni"	𐠜 = "no"	𐠝 = "nu"
𐠅 = "ja"			𐠈 = "jo"	
𐠲 = "wa"	𐠳 = "we"	𐠴 = "wi"	𐠵 = "wo"	
𐠨 = "sa"	𐠩 = "se"	𐠪 = "si"	𐠫 = "so"	𐠬 = "su"
𐠼 = "za"			𐠿 = "zo"	
𐠷 = "xa"	𐠸 = "xe"			

The Cypriot syllabary.

A handful of Cyriot characters looked like Linear B signs, a resemblance first pointed out in 1927 by the scholar A. E. Cowley:

Cypriot Syllabary		Linear B
𐠒	= "lo"	𐀫
𐠙	= "na"	𐀙

𐀞	= “pa”	𐠞
𐀡	= “po”	𐠡
𐀮	= “se”	𐠩
𐀲	= “ta”	𐠭
𐀴	= “ti”	𐠯
𐀵	= “to”	𐠰

The Cypriot syllabary was a millennium younger than Linear B, and, as investigators knew, a lot can happen to a script in a thousand years. Similar-looking scripts often record extremely dissimilar languages: English, Hungarian, and Vietnamese, for instance, are all written in versions of the Roman alphabet. Even in related languages, identical characters can have entirely different sound-values. In German, a close relative of English, the letter *w* is pronounced “v,” while the letter *v* is pronounced “f.” One has only to compare the American and German pronunciations of the word *Volkswagen* to take the point.

However tenuous the connection, the Cypriot script offered the only external clue the Linear B decipherers had to work with, and none could resist exploring it. In the early 1940s, Kober had tried plugging Cypriot sound-values into the Knossos inscriptions but, as she later wrote, “had no results.” Evans, too, had tried to exploit the clue. On one fragmentary tablet in particular, it seemed to rear up seductively:

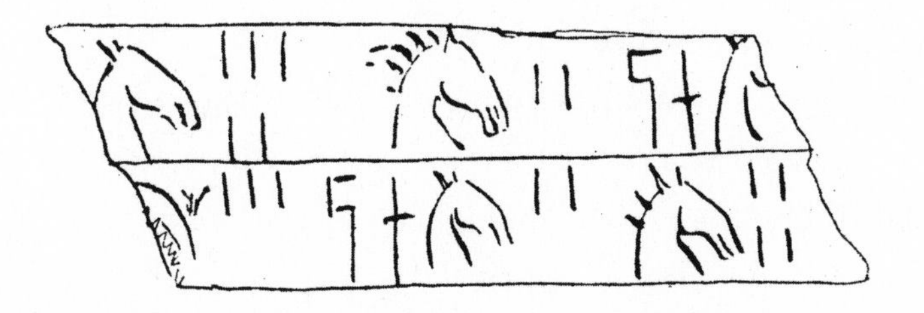

The fragment belonged to a tablet inventorying the horses of Knossos, and it appeared to count horses of two kinds: those with manes (shown at top center and bottom right), which were evidently full-grown horses, and those without (top left, top right, bottom center), evidently young horses. On the intact portion of the tablet, each maneless horse is preceded by a two-character Linear B word: 𐀡𐀫.

Evans tried substituting Cypriot values for those characters. What he got was "po-lo," which looked an awful lot like *pōlos*, the Classical Greek word for "foal." (The English word *foal* is a cognate, as is the name of the sport polo.) But championing Minoan supremacy to the last, he rejected the reading as mere coincidence—though one certain to be seized upon, as he wrote in a testy footnote in *The Palace of Minos,* by "those who believe that the Minoan Cretans were a Greek-speaking people."

Ventris, who also believed that Linear B recorded a non-Greek language, had remained suitably wary of this Cypriot clue. But in May 1952, after revisiting his earlier idea about Cretan place-names, he allowed himself to follow it. As an experiment, he plugged certain Cypriot sound-values into his third grid, shown on the next page.

He started by unpacking 𐀀𐀖𐀛𐀰, which he had previously rejected as "Amnisos," the Classical Greek name of the port of Knossos. From his analysis of the "pure vowel" signs, he was reasonably certain that 𐀀 stood for "a." He next turned to the Cypriot sign 𐠚, "na." If the Linear B sign 𐀙 had the same value, then he could insert "na" into the grid where Row C8 and Column V5 intersected—in the cell whose "phonetic coordi-

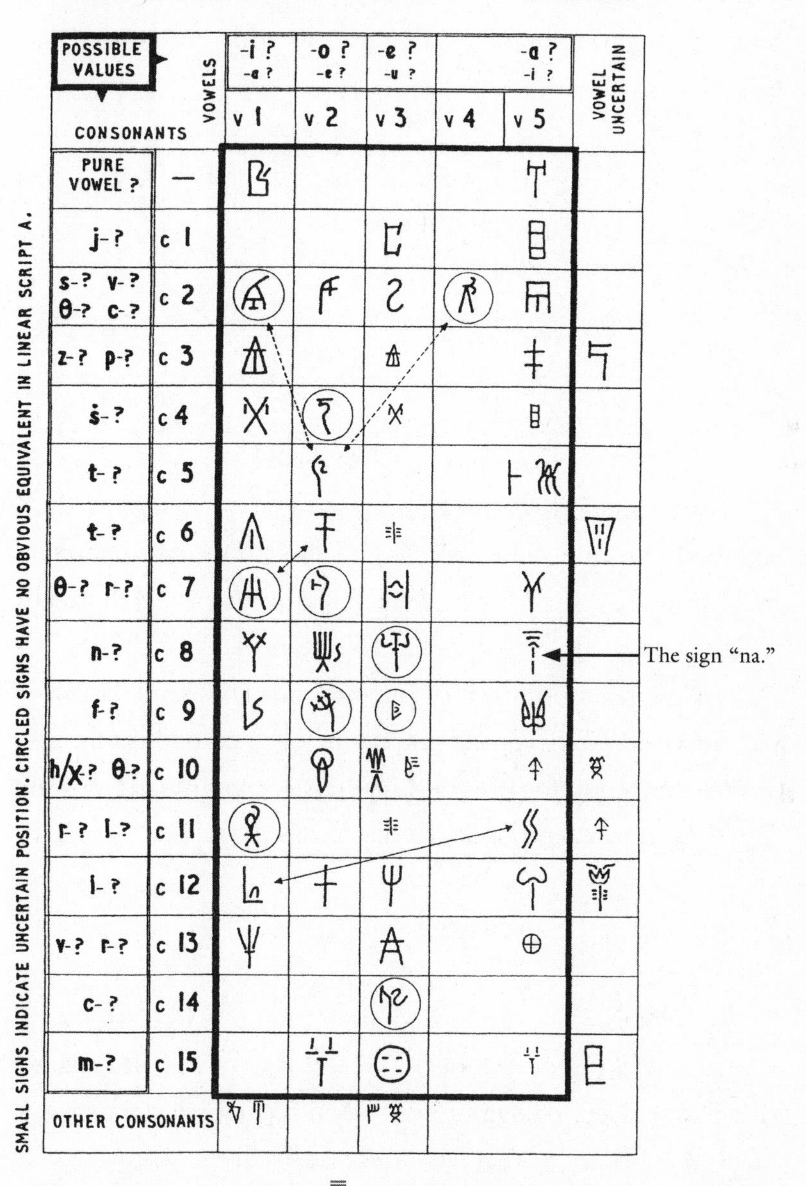

Ventris's third grid. The sign, “na,” placed at the intersection of Row C8 and Column V5 and indicated by an arrow, provided one of the first important clues to names inscribed on the tablets.

nates" were the consonant "n" and the vowel "a." (On his grid, Ventris draws the character as 𐀙, an acceptable variant form.) Simplified, the relevant portion of his grid now looked like this:

	V1 = i?	V2	V3	V4	V5 = a?
C6 = t?	𐀴				
C7					
C8 = n?	𐀛				𐀙
C9 = f?	𐀖				

Turning again to the Cypriot syllabary, Ventris tried assigning the sound-value "ti" to the Linear B sign 𐀴, analogous to Cypriot 𐠯. As it happened, he had already placed 𐀴 exactly where it ought to be: at the intersection of C6 ("t") and V1 ("i"). Now the grid's web of interdependencies truly began to pay dividends: His correct placement of 𐀴 automatically gave Ventris the value for 𐀛 ("ni") in the same column. The word 𐀀𐀖𐀛𐀰 so far was pronounced like this:

a- ______ - ni-______.

To Ventris, the word looked more and more like "aminiso," a syllabic spelling of "Amnisos." It was one of the place-names that had seemed to suggest itself when he first tried the experiment in February. If that were the case, then the word's second character, 𐀖, was "mi." (Ventris's initial placement of that sign as "fi" on the grid was incorrect.) Likewise, 𐀰 was "so."

Reading as 𐀀𐀖𐀛𐀰 as "a-mi-ni-so" immediately gave Ventris two more characters to plug into the grid. Those in turn gave him values for all the consonants in Row 7 ("s") and Row 9 ("m"), and for all the vowels in Column 2 ("o"):

	V1 = i	V2 = o	V3	V4	V5 = a
C6 = t	𐀴				
C7 = s		𐀰			
C8 = n	𐀛	𐀜			𐀙
C9 = m	𐀖				

He now turned to 𐀒𐀜𐀰, another "place-name" he had toyed with in February. From his revised grid, he knew that the third syllable was "so":

______ - ______ - so.

He had already placed the symbol 𐀜 at Row C8, Column V2, which gave it the value "no." If this were correct, the word now looked like this:

______ - no - so.

The incomplete word suggested a Cretan place-name—and not just any place-name but the single most important one on the island: Knossos, spelled syllabically as "ko-no-so." This let Ventris position 𐀒 correctly on the grid, where "k" and "o" meet:

	V1 = i	V2 = o	V3	V4	V5 = a
C6 = t	𐀴				
C7 = s		𐀰			
C8 = n	𐀛	𐀜			𐀙
C9 = m	𐀖				
C10 = k		𐀒			

Without the grid, identifying a name here or there would have yielded little additional return. With it, each new value "forced out" others, in precisely the kind of chain reaction Kober had foreseen. As a result, the third name Ventris had examined in February, 𐀶𐀪𐀰, could be read as "tu-ri-so," the Linear B spelling of Tulissos, also a Cretan town. As new sound-values burst from the grid, Ventris was able to read additional Cretan place-names in the Knossos inscriptions, including 𐀞𐀂𐀵 = "pa-i-to" (Phaistos) and 𐀬𐀑𐀵 = "ru-ki-to" (Luktos).

Once again, proper names had come to a decipherer's rescue. Ventris's great intuitive leap had proved correct: Kober's triplets represented the "alternative name-endings" of Cretan towns. That was his second significant contribution. Greek place-names alone, as Ventris knew, didn't prove that the language of the tablets was Greek. They could have been survivals carried over into Greek from an earlier, indigenous language—Ventris still held out for Etruscan. But in the coming weeks, as he worked the grid more deeply and new words emerged, he began to see something quite different from Etruscan. The process was like watching a photograph swim into being in a developers' tray, though what was coming to the surface didn't look much like Greek, either. Then again, of all the languages in the world that might be written with a syllabary, Greek is one of the worst candidates there is.

Greek screams for an alphabet. It is to the Greeks—who encountered the Phoenician alphabet at the start of the first millennium B.C., knew a good thing when they saw it, and improved upon it further—that we owe our own Roman alphabet, so handy for writing English. A "CV" syllabary like Linear B, on the other hand, which represents consonants and vowels in march-time alternation, is far less suited for writing Greek. Greek is a clatter of consonant clusters. (Consider, for example, the Classical Greek verb *epémfthēn,* "I was sent," fairly choked with contiguous consonants.) The language is also rife with side-by-side vowels, as at the beginning—and end—of the noun *oikía,* "house." For such words, too, Linear B is generally ill-equipped.

It had been understood since Evans's time that Linear B was an outgrowth of Linear A. Most investigators, including Evans and Ventris, thought the scripts wrote the same language. But what if they didn't? What if Linear B represented not the indigenous tongue of Minoan Crete but the language of later mainland colonizers? In that scenario, the invaders, unexposed to writing till they poured into Crete, seized the existing Minoan system for their own use. But if their native language was ill-suited to a syllabary, they would have needed to put the Cretan script through a set of compensating gyrations.

Though Ventris had held fast to his "Etruscan solution" since he was barely out of short pants, by the spring of 1952 he realized that he had to consider an alternative scenario: that the language of Linear B was not an indigenous Cretan tongue, but a Mycenaean import. That account would vindicate the few scholars, including Kober and Bennett, who thought the languages of Linear A and Linear B were different. It would also explain the presence of the B script on the mainland, brought back by victorious conquistadors for use at home.

Ventris began to wonder: Were the peculiar "spellings" of Linear B words the result of Cretan scribes having to force the square peg of the script into the round hole of a language for which it was never intended? Over the coming weeks, he worked out a set of "spelling rules" by which Linear B might have been used to write a non-Cretan tongue. The elucidation of these rules was his third major contribution. Among them were these:

One rule concerned the deletion of final consonants in Linear B words. Perhaps the most distinctive "fingerprint" of Classical Greek spelling is that words nearly always end in a vowel, or in one of a small set of consonants: "l," "m," "n," "r," or "s." When a "CV" syllabary is used to write such a language, it immediately runs into trouble, because it can't spell words that end in consonants. The scribe then has two options: He can insert a "dummy" vowel at the end of the word. (In such a system, the English word *cat* might be spelled "ca-t**a**.") Alternatively, he can delete final consonant altogether. (In this case, *cat* becomes "ca.")

Linear B chose the second alternative, and the word *po-lo,* so tartly dismissed by Evans, is a perfect example of the rule in action. The word was indeed *pōlos* ("foal"), impeccably respelled in Linear B. The problematic final "s" was simply lopped off, as the script's syllabic spelling rules dictated.

Strikingly, the Cypriot syllabary, when conscripted to spell Greek, made the opposite—though equally valid—choice. Instead of deleting final consonants, it tacked a “dummy” vowel on to the ends of words as needed: In Cypriot spelling, the Greek word *doulos* (“slave”), for instance, would have been written *do-we-lo-se,* with the final “e” serving as the dummy vowel. (In Linear B, by contrast, the same word is spelled *do-e-ro.*)

This difference in the handling of final consonants was the primary reason the Cypriot clue had been deemed useless for Linear B. As expected, Cypriot inscriptions in Greek contained a bevy of words ending in “se,” written by means of the character 𐠪. (These were words, like *doulos,* that actually ended in “s” in Greek.) If Linear B also wrote Greek, scholars reasoned, the tablets should be filled with words that ended with the corresponding character, 𐀮. But as the statistics of Kober, Ventris, and other investigators revealed, 𐀮 did not appear at the ends of Linear B words especially often. This reinforced the prevailing notion that the script wrote a non-Greek language.

Another “spelling rule” noted by Ventris concerned Linear B’s use of dummy vowels to break up Greek consonant clusters. The rule is at work in the very first syllable of “Knossos.” Written in Linear B, the word is “k**o**-no-so,” with the “o” of the first syllable intervening between the “k” and the “n.” (Note, too, that the final “s” has been deleted, exactly as Linear B’s spelling rules predict.)

Applying these rules and others, Ventris found he could read still more words. Among them were 𐀵𐀰 and 𐀵𐀭, the words for “total.” His grid gave them the values “to-so” and “to-sa”—perfect Linear B spellings of the Classical Greek words *tossoi* and *tossai,* the masculine and feminine plural forms of the word meaning “so much.” The Linear B signs for “boy” (𐀒𐀺) and “girl” (𐀒𐀷) came out as “ko-wo” and “ko-wa,” which suggested *kouros* and *korē,* the masculine and feminine singular forms of the Greek word for "child."

It was these words more than anything that forced Ventris to abandon his cherished "Etruscan solution" once and for all. "To-so" and "to-sa," along with "ko-wo" and "ko-wa," strongly suggested that the language of the tablets inflected its words for gender, much as French or Spanish does. Gender inflection is a hallmark of the Indo-European language family, to which languages like French, Spanish, German, Latin, and Greek belong. (What is more, *-o* and *-a* serve as masculine and feminine endings in a spate of Indo-European languages.) The presence of gender inflection in the tablets suggested a language other than Etruscan: Etruscan was non-Indo-European, and from everything scholars had gleaned about its grammar, it did not inflect words for gender.

In the coming weeks, Ventris pinpointed additional words, including 𐀡𐀕, "po-me" (like Classical Greek *poimēn,* "shepherd"); 𐀐𐀨𐀕𐀄, "ke-ra-me-u" (*kerameus,* "potter"; think of the English word *ceramics)*; 𐀏𐀐𐀸, "ka-ke-we" (*khalkēwes,* "bronze-smith"); and 𐀳𐀒𐀵𐀚, "te-ko-to-ne" (*tektones,* "carpenters"; compare English *tectonic,* which likewise has to do with the fitting together of things). Each new word meant new sound-values for the grid. These in turn generated other values.

On June 1, 1952, Ventris composed the twentieth and last in his series of Work Notes. This, his most famous Note, bears a title at once bold and tentative: "Are the Knossos and Pylos Tablets Written in Greek?" Just five pages long—far shorter than many previous Notes—Work Note 20 betrays the author's deep ambivalence about the solution that was massing unbidden before his eyes. "In the chains of deduction which spread out," Ventris wrote, "we may, I believe, initially strike words and forms which force us to ask ourselves whether we are not, after all, dealing with a *Greek* dialect."

Then, in the very next paragraph, he retreats, writing, "These may well turn out to be a hallucination." He then continues: "If we were to toy with the idea of an early Greek dialect, we should have to assume that a Greek ruling class, appearances to the contrary, established itself at Knossos as early as 1450, and that the new Linear B was adapted from the indigenous syllabary in order to write Greek."

Over the next several pages, Ventris lays out the results of his place-name experiments before closing with a final retreat: "If pursued, I suspect that this line of decipherment would sooner or later come to an impasse, or dissipate itself in absurdities; and that it would be necessary to revert to the hypothesis of an indigenous, non-Indo-European language."

But over the next few days, try though he might to suppress it, Greek kept asserting itself. One night in early June, he found he could no longer fend it off. That night, the journalist Prue Smith and her husband had been invited to dine at the Ventrises' flat; Smith's husband, also an architect, was a colleague. When they arrived, as she recounts in her memoir, Michael Ventris was nowhere to be seen:

> Lois Ventris, whom we always called Betts, talked amiably, apologising at fairly frequent intervals for her husband's absence—he was in his study, she said, and would come as soon as he could. We grew a little hungry, and a little drunk on the pre-prandial sherry, and Betts a little anxious and embarrassed. After what seemed a very long time Michael

> burst into the room, full of apologies but even more full of excitement. "I know it, I *know* it," he said, "I am certain of it——." I thought he must have confirmed an earlier idea that the language was Etruscan. But what he had proved, of course, was that the language was an early form of Greek . . . recorded on the earliest known documents of European civilisation.

After spending half his life on the problem, Michael Ventris, a month shy of his thirtieth birthday, had solved the riddle of Linear B. But before long, just as the script itself had, the consequences of being its decipherer would start to consume him.

12

SOLUTION, DISSOLUTION

PRUE SMITH WAS A RADIO producer at the BBC. She had also studied classics. As Ventris, in high excitement, explained his discovery that tipsy night in his flat, she knew she was witnessing something extraordinary: Her friend had just solved one of the most intractable problems in history.

On July 1, 1952, after reworking his script several times, Ventris took the microphone at BBC Radio. The broadcast, as Andrew Robinson notes, is the only record we have of his voice: high, light, cultured, melodious, with "a curious combination of firmness and diffidence, reflecting the brilliant but still unproven nature of his discovery." The program was called "Deciphering Europe's Earliest Scripts," and it would prove transformative—for Ventris, for scholarship, and for the collective understanding of European history. Ventris said:

> For half a century, [the] Knossos tablets have represented our main evidence for Minoan writing, and many people—classical scholars and archaeologists as well as dilettanti of all kinds—have been fascinated by the problems of deci-

> phering them. Until now they have all been uniformly unsuccessful. . . .
>
> With the almost simultaneous publication of the Knossos and Pylos tablets, all the existing Minoan Linear Script material is now available for study, and the race to decipher it has begun in earnest. . . .
>
> For a long time I . . . thought that Etruscan might afford the clue we were looking for. But during the last few weeks, I have come to the conclusion that the Knossos and Pylos tablets must, after all, be written in Greek—a difficult and archaic Greek, seeing that it is 500 years older than Homer and written in a rather abbreviated form, but Greek nevertheless. . . . Once I made this assumption, most of the peculiarities of the language and spelling which had puzzled me seemed to find a logical explanation. . . .

What Ventris disclosed in the broadcast was breathtaking: Well before the Greek language was thought to exist, the first Greek-speaking people, unruly and unlettered, swarmed into Crete. There, they appropriated one of the indigenous writing systems—Linear A—that had flourished on the island for generations. The Knossos tablets and their later counterparts at Pylos were written, against all expectations, in a very early version of the language of Plato and Socrates, set down centuries before the advent of the Greek alphabet. Chronologically, the Greek dialect they contained was as distant from Classical Greek as the Anglo-Saxon of *Beowulf* is from Shakespeare.

Ventris's announcement brought him a measure of renown in Britain. It also brought him critics—the dubious, the envious, the bewildered, and the vitriolic—who simply could not credit that a mere amateur had solved one of the greatest riddles in Western letters. Even Bennett and Myres remained unpersuaded at first. But the deepest doubts came from Ventris himself. In his seminal Work Note 20, which he had described as "a frivolous digression," he had taken pains to call attention to features of the script that appeared incompatible with Greek. In the weeks after the decipherment, as his private correspondence makes plain, he continued to fear that his own solution was the product of smoke and mirrors.

Ventris gained a valuable ally in John Chadwick, a classics professor at Cambridge University. Chadwick was interested in the Aegean scripts but was not part of the circle of scholars actively working on them; neither Kober nor Ventris had been in correspondence with him. But he was a specialist in early Greek dialects, and a former wartime cryptographer. Ventris's broadcast left him intrigued if skeptical. He got in touch with Myres, who handed him Ventris's Work Notes. As Chadwick later described the visit:

> He sat as usual in his canvas chair at a great desk, his legs wrapped in a rug. He was too infirm to move much, and he motioned me to a chair. "Mm, Ventris," he said in answer to my question, "he's a young architect." As Myres at that time was himself eighty-two, I wondered if "young" meant less than sixty. "Here's his stuff," he went on, "I don't know what to make of it. I'm not a philologist.". . .
>
> I approached the matter very cautiously, for impressed as I had been by the broadcast, I had a horrid feeling the Greek would turn out to be only vague resemblances to Greek words . . . and wrong for the sort of dialect we expected.

Chadwick began plugging Ventris's sound-values into the published inscriptions, and before long he was a convert. He also discovered additional Greek words on the tablets that had escaped Ventris's notice. "I think we must accept the fact," he told Myres, "that a new chapter in Greek history . . . is about to be written."

In mid-July, Chadwick wrote to Ventris directly. "Dear Dr. Ventris," he began, a salutation that suggests he believed Ventris to be an academic. "Let me first offer my congratulations on having solved the Minoan problem; it is a magnificent achievement and you are yet only at the beginning of your triumph. . . . Ever since hearing your talk on the wireless I have been most excited, and when Sir John showed me your provisional list of identifications last Monday I set to work at once to verify your discovery."

In closing, Chadwick made a gracious offer: "If there is anything a mere philologist can do," he wrote, "please let me know." Whether he realized it or not, he had just thrown Ventris a lifeline, and Ventris grasped it gratefully: Chadwick had a deep knowledge of pre-Classical Greek, as well as the academic bona fides Ventris lacked.

"Frankly at the moment I feel rather in need of moral support," Ventris replied. "The whole issue is getting to the stage where a lot of people will be looking at it very skeptically, and I am conscious there's a lot which so far can't be very satisfactorily explained. . . . I've been feeling the need of a 'mere philologist' to keep me on the right lines. . . . It would be extremely useful to me if I could count on your help."

Chadwick was able to allay some of his fears right away. One involved the word "the," or rather the lack of it, on the tablets.

The Classical Greek texts Ventris had studied as a schoolboy teemed with the definite article in its welter of forms: masculine, feminine, and neuter genders, singular and plural, in each of five grammatical cases. But he couldn't find "the" anywhere on the Linear B tablets, and that worried him. In fact, Chadwick assured him, that was precisely to be expected in Bronze Age Greek. Even five hundred years later, in the Greek of Homer's time, "the" was a rarity, taking firm root in the language only in the Classical Age.

In the summer of 1952, the two men embarked on a close collaboration that would last to the end of Ventris's life, with Chadwick, as Robinson observed, playing the dogged Watson to Ventris's inspired Holmes. An article they would write together, "Evidence for Greek Dialect in the Mycenaean Archives," brought the decipherment to a wide scholarly audience. Another, published in the British journal *Antiquity,* brought it to a wide popular one. They also began work on a massive book, *Documents in Mycenaean Greek,* which would become the bible of Linear B studies, though Ventris would not live to see it in print.

Despite the sense of security the collaboration should have brought him, Ventris "had attacks of cold feet that summer," as Chadwick recalled in *The Decipherment of Linear B.* His "Greek solution" continued to cause him anguish. "Every other day I get so doubtful about the whole thing that I'd almost rather it was someone else's," Ventris wrote Chadwick at the end of July.

Ventris's doubts crept into the prestigious public lectures he was increasingly asked to give. Before one talk, at Oxford, Chadwick had to counsel him not to shy away from his own solution. "I feel it would be appropriate . . . to be a little more definite in asserting the language to be Greek," Chadwick told him. "A proper intellectual humility is a good thing, but (especially at Oxford) it may be mistaken for diffidence."

As Ventris was well aware, many still viewed the decipherment with suspicion. Evans's hold over Classical archaeology had been iron-fisted, and even a decade after his death, many scholars still believed that the language of Linear B could not possibly be Greek. Their criticism centered in part on the seeming malleability of Ventris's spelling rules, which permitted the deletion of consonants, the insertion of vowels, and other alterations at strategic points in Linear B words. The character 𐀡, for instance, as Ventris acknowledged, could stand not only for "po" (as in the word "po-lo"), but also for "pho," "poi," "phoi," "pol," "phol," "pom," "phom," "pon," "phon," "por," "phor," "pos," "phos," and many other syllables. With desire, imagination, and mercurial spelling rules, the critics argued, a decipherer could mold the inscriptions to mimic practically any language. Hrozný had done precisely this when he "read" the Knossos texts as a form of Hittite. Ventris, his antagonists believed, had done likewise.

In the spring of 1953, vindication came from an improbable source: the American archaeologist Carl Blegen, previously so unforthcoming with the tablets he had unearthed at Pylos. Digging there again in 1952, Blegen found more tablets, and it was one of these that would help Ventris silence his critics. The tablet, known officially by the unromantic name P641, was a handsome thing. Long, slender, and nearly intact, it was inscribed with three lines of text. Like most Linear B tablets, it was an inventory, in this case of pots, jars, and other vessels:

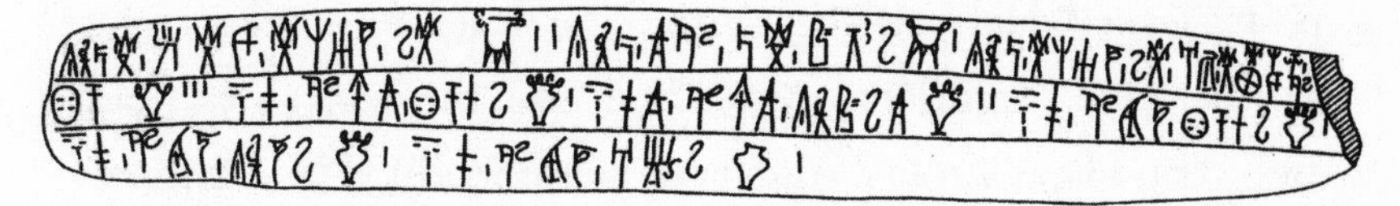

Pylos Tablet 641, known as the "Tripod" tablet, would confirm Michael Ventris's decipherment of Linear B. This copy was hand-drawn by Ventris himself.

After plugging some of Ventris's sound-values into the inscription, Blegen knew immediately that this tablet confirmed the decipherment. In May 1953, after the tablet had been cleaned, catalogued, and copied, he mailed a transcription to Ventris in a striking act of collegiality. P641 wasn't a bilingual, but it was the next best thing: a list containing names of objects accompanied by logograms for those objects. In a sense, it *was* a bilingual, inscribed in Greek and the universal language of pictures. It would prove to be the decipherment's crowning confirmation.

The tablet's opening phrase consisted of a four-word inscription, followed by a picture of a three-legged cauldron and the number 2. It looked like this:

When Ventris's sound-values were plugged in, the phrase read this way:

ti-ri-po-de ai-ke-u ke-re-si-jo we-ke

Blegen recognized the first word, "ti-ri-po-de," as the Linear B spelling of the early Greek word meaning "two tripods"—the "dual" number of the Classical Greek word *tripos,* "three feet." Translated, the phrase on the tablet described two "tripod cauldron(s) of Cretan workmanship," and a three-footed cauldron was precisely what the pictogram showed. "Is coincidence excluded?" he wrote to Ventris in May.

The rest of the tablet was equally exciting. One phrase, 𐀇𐀞𐀁 𐄀 𐀕𐀿𐀁 𐄀 𐀴𐀪𐀃𐀺𐀁, could be transliterated *dipae mezoe tiriowee,* "larger-sized goblets with three handles." A picture of a three-handled vessel appeared at the end of the phrase. Another, 𐀇𐀞 𐄀 𐀕𐀿𐀁 𐄀 𐀤𐀵𐀫𐀸, *dipa mezoe qetorowe* ("larger-sized goblet with four handles"), was followed by a picture of a four-handled vessel. The word *dipa,* "goblet" (*dipae* is the plural), recurs, only slightly altered, in Homer: In the *Iliad,* Nestor, King of Pylos, has a cup so large that when filled, it can scarcely be lifted. It is called a *depas.* Other phrases on P641 described other kinds of vessels ("smaller-sized goblet with three handles," "smaller-sized goblet without a handle"), each accompanied by the corresponding logogram.

On receiving Blegen's transcription, Ventris telephoned Chadwick in Cambridge. Making a long-distance call was noteworthy in those days; more noteworthy still, as Chadwick recalled, was the "great state of excitement" in Ventris's voice. "He rarely showed signs of emotion," Chadwick wrote, "but for him this was a dramatic moment."

There could be little doubt now that coincidence was excluded: The tablets were unquestionably written in Greek. Even Bennett, until then a holdout, wrote to Ventris, "Looks hard to beat!" after seeing the "Tripod" tablet, as P641 quickly became known. Some time afterward, addressing an international classical conference in Copenhagen, Ventris showed a slide of the

"Tripod" tablet. On seeing it, the audience burst into spontaneous applause. (It "went off all right" was how he described the talk to Chadwick afterward.)

Ventris was in constant demand now, giving lectures all over the world, often in the host country's native language. He spoke before the king of Sweden. He spoke at Oxford. He spoke at Cambridge. In June 1953 he lectured at Burlington House in London, coming full circle to the place where as a boy he had excitedly queried Arthur Evans. The next day, the *Times* of London carried an account of Ventris's lecture on page 1, opposite an article by Edmund Hillary about his summit of Mount Everest the month before. As a result, the decipherment became known as "the Everest of Greek archaeology," a description whose Olympian hyperbole Ventris found galling.

But in fact, as the classicist Maurice Pope has written, Ventris's accomplishment is in many ways the more spectacular of the two: "Whichever is regarded as the greatest personal feat or the most important in its consequences, there can be no question which of them belongs to the rarest category of achievement," he says in *The Story of Archaeological Decipherment*. "People in other societies have climbed mountains. . . . But the recovery of the key to an extinct writing system is a thing which has never been attempted, let alone accomplished, by anybody except in the last two or three centuries of our own civilization."

In August 1953, the joint article by Ventris and Chadwick, "Evidence for Greek Dialect in the Mycenaean Archives," appeared in the *Journal of Hellenic Studies,* a major British academic publication. In it, they proposed sound-values for more than sixty signs and gave long lists of vocabulary: place-names;

men's and women's names; names of occupations like priest, armorer, physician, cook, and baker; and names of commodities like chariots, grain, and pigs.

Most telling of all was the authors' decision to abandon the term *Minoan,* so beloved by Arthur Evans, and call all Linear B inscriptions—be they Cretan or mainland—Mycenaean. What they were saying, as Chadwick later wrote, was that "the label 'Minoan' had been out of date as far as Linear B was concerned since 1939. . . . With our conviction that Linear B contained Greek went the irresistible conclusion that Knossos in [that] period formed part of the Mycenaean world." Evans's beloved Minoans, with their high style and sophisticated civilization, would have to wait for Linear A to be unraveled for their language to be revealed. (The script, used between about 1750 and 1450 B.C., remains undeciphered to this day. It may forever remain so: there is simply not enough text from which to work.)

In the spring of 1954, an article by Ventris about the "Tripod" tablet, "King Nestor's Four-Handled Cups," was published in the American journal *Archaeology,* helping to bring the decipherment to an international readership. Ventris was truly world-famous now, showered with interview requests, and with honors. In 1955, the young Queen Elizabeth appointed him to the Order of the British Empire. "Offers to join the academic world . . . were now his for the asking," Robinson writes. But Ventris, no doubt keenly conscious of his lack of a university education, and his mere three years of schoolboy Greek, turned them all down.

Like Kober, Ventris never saw the decipherment as a contest, and was ill at ease in the spotlight. Despite the vindication of the "Tripod" tablet, and despite his growing renown, he was still racked with worry about his solution. "After the *Times* article I had a letter from a crank," he wrote Chadwick in the summer of 1953. "The trouble is that, ridiculous as his ideas are, one always has the uneasy feeling of 'there, but for the grace of God . . .'; and one's worst nightmare is that one has oneself been a victim of a similar delusion."

In the summer of 1953, Ventris, Lois, and the children moved into a Modernist house he had designed for them in Hampstead, an exclusive part of London. But he was otherwise uninvolved with his family. His abrupt, youthful marriage to Lois appeared to have run its course; given Ventris's seeming inability to make deep connections with other people, it is surprising that it endured as long as it did, for Lois turned out to share neither his profound intellectualism nor his passion for Linear B. As Robinson writes:

> By 1956, after fourteen years of marriage, the Ventrises had drifted quite far apart. A friend from the Ministry of Education, Edward Samuel . . . took several enjoyable holidays with [them] in the 1950s. . . . One day, Lois told him candidly that the reason he was invited along was because otherwise she and Michael would simply run out of conversation.

Nor did Ventris seem to have forged strong bonds with his children. "I didn't find him easy to live with," his daughter, Tessa, then in her fifties, said in the documentary film *A Very English Genius.* "I think his relationships with people were on his own terms. . . . If he wanted to be amusing and charming, he would be amusing and charming. And if he wanted to go away and work, he would go away and work. . . . I admired him, but don't think I liked him." She added: "That was probably just jealousy, because he got all the attention."

In July 1955, Ventris and Chadwick completed the manuscript of their 450-page book, *Documents in Mycenaean Greek.* Though work on the script still beckoned—thanks to the decipherment, Ventris had lectures to give, international meetings to attend, journal articles to write, and longs lists of Linear B vocabulary to type up for fellow scholars on a special typewriter called a Varityper—he soon tried to bow out of the field one more time.

"I shan't be able to devote time to any other major commitments," he wrote Chadwick in December. "Once the two present pieces of typing are done, there's not much for me to do anyway."

THIS TIME, AT LEAST, Ventris was bowing out for a good reason: He'd been given the chance to return to architecture. *Architects' Journal,* the field's principal publication in Britain, had just inaugurated a fellowship: a thousand pounds to a member of the profession to support a year's research on an architectural topic. Its first recipient—a signal honor—was Ventris, who would begin his research project at the start of 1956. The topic he chose was "Information for the Architect: What Does He Need and Where Will It Come From?"

At the time, British architects were working more or less in isolation. There was no universally available information on materials or methods, and members of the profession depended on word of mouth (and their time-honored ways of doing things) to ply their trade. "There were almost no books on architectural design, explaining how to design a school, a hospital or a factory," Robinson writes. "The typical architects' design office did not even have a bookshelf." In many ways, the predicament of architects mirrored that of midcentury Linear B scholars: both entailed isolation, ignorance of far-flung colleagues' work, and the lack of reliable guides to practicing one's craft.

What Ventris sought to do was to create a large-scale, architectural version of his Linear B questionnaire of 1949, from which he could produce a state-of-the-field dossier much like his Mid-Century Report. Just as he (and Kober before him) had dreamed of setting up a central clearing house for Linear B, he now envisioned one for architectural information of all kinds. "One might ring up the information centre to ask for any information, say, on *aluminium schools* in Australia," Ventris wrote in a preliminary report. "The information officer would operate a keyboard with the . . . numbers for *aluminium, school* and *Australia*; the microcards comprising the complete information of the centre would be sorted for those sharing these codings; the selected cards would have their articles transmitted electronically to a view or printer at the subscriber's desk." What he was proposing was essentially a high-tech version of Kober's cigarette-carton database.

But the project soon paled, and Ventris began to feel like an outsider in this field, too. In the wake of the decipherment, with all its attendant publicity, he had become far better known for his work on Linear B. In a sense, he had lost two careers: He was conducting research *about* architecture but not actually practicing architecture; nor was he a decipherer anymore. He had been drawn to Linear B by the pure mathematics of decipherment, with its crystalline clarity and relative certainties. Its humanities aspects—what could be gleaned from the tablets about a long-ago civilization—interested him far less. Though his work gave birth to an entire branch of ancient history known as Mycenology, Ventris would eventually tell Emmett Bennett "that he himself saw no future in Linear B," as Robinson writes. By mid-1956, Ventris had become an interloper twice over, with a foot in each of two worlds but secure bearings in neither.

In June, Ventris turned in the first half of his fellowship report to *Architects' Journal.* He would never write the second: The work, as he later put it, had become "cold and dull." Oliver Cox, his architecture-school classmate, "was aware," Robinson writes, "that his gifted friend was constantly depressed that spring and summer."

On August 22, Ventris sat down and composed what Robinson calls an "extraordinary, shocking, abject, private letter" to the editor of *Architects' Journal.* Written by hand in his impeccable printlike script, it renounced the remainder of the fellowship. It recalls Ventris's letter of abdication to Myres in 1948, but is even more raw, more self-lacerating, and more final:

> *I have had a couple of weeks abroad, and had a chance to get into perspective the hash that I've been making of your Fellowship; I've come to the conclusion that it's quite unrealistic for me to pretend to you or to myself that I'm going to be able to finish off the work in the way that it should be done. I'm afraid I must ask you again, as I have done since April, to devise some formula, however humiliating to myself, for relieving me of the second part of the task.*

I am mortally ashamed of the waste of time and energy that this false start means for those who have been associated with the Fellowship. It would be easy to say that the "information" subject was a dangerous research subject, and that it was risky to pick on me to do it; but the fault lies in me, and I know how worthwhile the job would be if one could do it well. The peculiarities of mind and personality which seemed to make me suited for the Fellowship have turned on me and made me deeply doubt the value of both my vaunted intelligence and to a large extent that of life itself. . . .

The money I've so far received from you on account will have to be paid back apart from any value that you may put on the work I have already turned in: perhaps that can stand as some sort of contribution, though I know only too well how cold and dull it all is. As for explaining to AJ *readers why the second half of the programme has flopped, you'd be justified in writing me off in a way that will make it difficult to hold up my head in the ranks of architects again, and bring pain to my family. All I can ask you is to temper your justifiable anger with a little compassion.*

Yours,

Michael Ventris

Very late at night on September 5, 1956, Ventris left home alone in his car. He apparently told his family that he was going to retrieve his wallet, which he had left at the home of Lois's parents. But what wallet cannot wait until morning?

Just after midnight on September 6, Ventris, north of London, pulled at high speed into a rest area off the main road. He collided with a parked truck and was killed instantly. His magnum opus, *Documents in Mycenaean Greek,* written with Chadwick, was published by Cambridge University Press a few weeks later. It remains a seminal text in Mycenaean studies.

At the coroner's inquest, the death of Ventris, thirty-four, was ruled an accident. But the question of whether he took his own life is debated by classicists to this day. His family has long maintained that his death was accidental. "I don't think he committed suicide," Tessa Ventris said in *A Very English Genius.* "He was much too positive for that." She said she thought her father had a heart attack and blacked out at the wheel. Heart disease may well have run in the family: Ventris's son, Nikki, died of a heart attack in 1984, in his early forties.

In the end, it is impossible to know exactly what happened. Perhaps it does not matter. For Ventris's achievement—startling, monumental, and incontrovertible—will endure down the ages.

AND SO THE STORY ENDS, bracketed by two architects: Daedalus, who built the Minoan labyrinth, and Ventris, who found the thread that unraveled the tangle of writing unearthed there. But however Olympian his accomplishment, it should not be forgotten that Ventris attained it by standing on the small, round shoulders of an unheralded American giant, a fact he acknowledged less conspicuously than he might have.

In his pathbreaking BBC announcement of July 1952, when he unveiled his discovery to the world, Ventris took pains to credit Bennett and Myres by name. But he did not mention Kober, whose syllabic grid, with its carefully worked-out abstract values, was the foundation stone of his decipherment.

"It may be interesting to discuss just how one sets about

a job like this," Ventris explained on the radio that day. "It is often alleged to be impossible to decipher a set of inscriptions where both the writing and the language are unknown quantities, and where there is no bilingual to help us. But provided there is enough material to work on, the situation is not hopeless at all. It simply means that, instead of a mechanical piece of decoding, a rather more subtle process of deduction has to be undertaken. It is rather like doing a crossword puzzle on which the positions of the black squares have not been printed for you."

But it had been Kober, after all, who supplied those first black squares—enough of them to let the puzzle be solved. It was she who, after poring for years over the snarl of symbols and cutting out tens of thousands of cards, identified the language of Linear B as inflected. That was the decipherment's essential first step. It was she who put her finger on the singular interaction between an inflected language and a syllabic script, pinpointing the critical "bridging" character. That was the second step. And, in the third step—her masterstroke—it was she who realized that it was possible to plot the relationships among characters in the abstract, drawing up the very grid on which Ventris later built.

It was also she who had determined at the start that the only hope of cracking the code lay in hunting down and analyzing internal patterns in the script, without speculating on either the underlying language or the sound-value of any character. And that was as essential to the decipherment as anything.

In a scholarly lecture he gave in 1954, published posthumously in 1958, Ventris did credit Kober at some length, speaking of her "series of fundamental articles" that constituted "the first systematic programme of analysis and research of the Linear B documents." But it was too little, too late, and for too minute an audience.

The question beckons: Could Kober, given more time, have solved the riddle of the script? Andrew Robinson thinks not, writing that she "was probably too restrained a scholar to have 'cracked' Linear B." I am not so certain. Granted, I have a brief for Kober, just as Robinson did for Ventris. But let us, as she might have said, consider the facts, which her archived correspondence lets us do more fully than at any time in the past. Kober was cautious, yes, but her private papers reveal a genuine willingness to experiment—at her dining table, at least, if not yet in print. We know, for instance, that very early on, she allowed herself to play with the Cypriot syllabary, plugging its sound-values into the Knossos inscriptions. The fact that she "had no results" is hardly surprising, given the paucity of Linear B inscriptions then available and the corresponding poverty of knowledge about the script.

Her careful ten-cell grid of 1948, as Maurice Pope points out, made possible every stage of the decipherment that followed. "Every one of Kober's inferences . . . has been supported by Ventris' decipherment," he wrote in *The Story of Archaeological Decipherment.* "Kober's method of trying first to establish the interrelationship of the phonetic values of particular signs on an abstract level was as unique as it was fruitful."

She *was* wrong about one thing, though only partly. But it was this error that ultimately let Ventris slip past her to make the intuitive leap that allowed him to decipher the script.

Kober's mistake concerned inflection. She was right that the language of Linear B inflected its words for gender, as the two forms of the word "total," and the words "boy" and "girl" showed. She was also right that it inflected its words for case—the who-does-what-to-whom of sentences. But she was wrong about precisely *which* words those actually were.

Her error centered on the sets of words that Ventris called "Kober's triplets." They included this set, from her 1948 article:

As Kober realized, the "triplets" *did* represent a series of word endings—, and —attached to the same stem. But the endings in question were not precisely the kind she had supposed, and that turned out to make all the difference.

Many languages, including Greek and English, use two different types of word endings, called "grammatical" and "derivational." Grammatical endings, which include case inflections, give a sentence its syntax. Derivational endings, by contrast, create bigger words from littler ones. In English, grammatical endings are rare—limited to the lonely final *-s* on third-person-singular verbs; the past-tense suffix, *-ed*; and a half-dozen others. But derivational endings are legion. They include *-ity,* which turns an adjective into the corresponding noun *(scarce, scarcity)*; *-ing,* which turns verbs into participles *(swing, swinging)*; *-able,* which changes a verb into an adjective meaning "capable of being . . ." *(sing, singable)*; and many more.

Kober thought her triplets represented inflection. Ventris thought they represented derivation—"alternative name-endings," he called them—and he was right. Which type of endings the triplets turned out to be made no difference to Kober's essential theory: Either way, the "bridging" characters function precisely as she described them. But recognizing the triplets as having *derivational* rather than *grammatical* endings is what let Ventris surpass the ground she had gained.

In his great intuitive leap, Ventris guessed that the "alternative name-endings" in the triplets were derivational variants of place-names—variants along the lines of *Brooklyn/Brooklynite/Brooklynese.* He found that one such word, 𐀒𐀜𐀰 (the last word in the triplets above), could be transliterated as *ko-no-so,* the name for Knossos itself. The word just above it, 𐀒𐀜𐀯𐀍, clearly had the same stem. It turned out to be *ko-no-si-jo,* "men of Knossos." The topmost word, 𐀒𐀜𐀯𐀊, was *ko-no-si-ja,* "women of Knossos."

The other triplets, he found, behaved similarly:

𐀀𐀖𐀛𐀰 = *a-mi-ni-so* ("Amnisos")
𐀀𐀖𐀛𐀯𐀍 = *a-mi-ni-si-jo* ("men of Amnisos")
𐀀𐀖𐀛𐀯𐀊 = *a-mi-ni-si-ja* ("women of Amnisos")

Proper names like these turned out to be central to Ventris's decipherment, as they have been to many of the great decipherments in history. But in her private papers, there is evidence that as early as 1947, Kober was willing to consider the same idea. The evidence comes from an unlikely source: the bug man, William T. M. Forbes, professional entomologist and amateur etymologist. In a letter to Kober dated May 1, 1947, Forbes floated an idea he had proposed to her many times before: that on certain tablets, at least, some of the inflected words in her paradigms represented the names of Cretan towns. In the margin of his letter, in her unmistakable pedagogical hand, Kober penciled the annotation, "Agree—place-names."

If her teaching load had not been so great, if her Guggenheim Fellowship had been renewed, if she had been hired at Penn after all, if Myres had not saddled her with a crushing secretarial load, if her champion John Franklin Daniel had lived—if *she* had lived—it is entirely possible that Alice Kober would have solved the riddle of Linear B. Among her papers in the archives of the University of Texas is an undated notebook in which she constructed a phonetic grid containing more than twenty Linear B characters—more than twice the number of her published grid. She never published this larger grid, nor did she assign sound-values to any character on it. But as Ventris's decipherment would show, her relative placement of every character was correct. She was clearly poised to make headway, if only she had been given time.

What is beyond doubt is this: Without Kober's work, Linear B would never have been unraveled as soon as it was, if ever. Her deep intellect, her single-minded resolve, and her ferocious rationalism made it possible to recapture the vanished key to the script, the earliest Greek writing of all.

EPILOGUE

MR. X AND MR. Y

THERE ARE NO GRAND NARRATIVES lurking in Linear B—no epic poems, no romances, no tales of gods and their derring-do. Arthur Evans knew as much from the start, as did every serious investigator after him. They were all aware, as Alice Kober reminded her Hunter College audience that June evening in 1946, that "we may only find out that Mr. X delivered a hundred cattle to Mr. Y on the tenth of June, 1400 B.C." And that, of course, is precisely what they did find: records of crops harvested, goods produced, animals tended, and gifts offered up to the gods.

As a result, some observers have deemed the postdecipherment tablets dull and dispiriting. In *The Man Who Deciphered Linear B,* Andrew Robinson writes, "As for what the humanities—archaeologists, historians, literary scholars and others—have learnt from the decipherment since Ventris's death, the answer is, honestly speaking, a little disappointing, set beside the artistic treasures of Troy, Mycenae and Knossos."

But the tablets are unrivaled treasures for the light they shine on Mr. X and Mr. Y and their Bronze Age world—the world of Odysseus, Nestor, and Agamemnon. The great American newspaperman Murray Kempton, remarking sagely on the difference between criminal and civil proceedings, once wrote: "The Criminal Courts can only tell us the way some of our sisters and brothers steal or kill or die. But the Civil Courts tell us the way all of us live." The same is true of the tablets—the civil documents of the first Greeks.

In the course of three millennia, the Linear B tablets passed from complete readability to complete obscurity and, against all odds, back to readability again. They reveal much about who the Mycenaeans were, from aristocrats through artisans and tradesmen and down to slaves. Though scholars continue to debate the precise interpretation of particular tablets, the Linear B archives as a whole disclose the day-to-day workings of a going civilization three thousand years distant, including, as the Mycenologist Cynthia W. Shelmerdine has written, "the movement of goods . . . , the status of land and animal holdings, the manufacture and repair of various kinds of equipment, and the personnel needed to carry out all the business of a Mycenaean state."

The members of that state were flesh-and-blood men and women, as the tablets clearly show. Their account books, set in clay and baked in unintended fire, tell us what they sowed and reaped, what they ate and drank, the names of the gods they worshipped (with members of the Greek pantheon standing shoulder to shoulder with strange, pre-Greek deities), how they earned their keep, how they passed their time, how they defended themselves and made war. We even know their names, some of them names of exquisite nobility, others names one wouldn't wish on a dog.

* * *

"ALMOST ALL PARTS of Greece became Mycenaeanized," the scholar J. L. García Ramón has written; their combined population, spread over more than 150 communities, was about fifty thousand. The principal Mycenaean kingdoms were these:

There was Knossos, where invading Greeks took over the existing Minoan palace in about 1500 B.C. and held sway for a century or less, until in some unknown catastrophe the palace burned to the ground. There was Pylos, on the Greek mainland, home to the Palace of Nestor. The tablets there are younger than those of Knossos, and indeed, Mycenaean civilization managed to hang on there about two hundred years longer before it, too, was extinguished.

Another mainland kingdom was Mycenae itself, excavated by Schliemann in the 1870s. This was the kingdom that had propelled Evans on his quest for writing, for he was certain that so fine a civilization could scarcely have done without it. As it turned out, he was right. In 1952, about forty Linear B tablets were uncovered at Mycenae, not far from where Schliemann had dug. So here, too, as Evans had long suspected, an advanced, literate Bronze Age kingdom had flourished. Elsewhere on the mainland, the script has surfaced at Tiryns and at Thebes. Isolated finds continue to be made; as recently as 2010, a small piece of a Linear B tablet, preserved by a fire in an ancient refuse pit, with text pertaining to manufacturing of some kind, was discovered near the village of Iklaina, in southwest Greece.

The tablets show the twilight of the kingdoms. Normally, scribes pulverized their written records at the end of each year. The granules of unfired clay were mixed with water, and from this paste the next year's tablets were formed, a practice that conserved both clay and storage space. Scholars have conjectured that before each crop of tablets was destroyed, the year's records were transferred to a more permanent medium—permanent for its day, anyhow—like ink on parchment. But since any trace of these materials would have vanished long ago, whether the Mycenaeans actually did this can never be conclusively known.

So what we have, then (and all we will ever have), are the records of the final year of each palatial center before some cataclysm—invasion, earthquake, lightning strike—and the subsequent fire reduced the Mycenaean Age to ash.

THESE WERE THE PEOPLE of the kingdoms: "Mycenaean state bureaucracy," Cynthia W. Shelmerdine writes, "was highly centralized, and authority rested in the hands of a hierarchy of officials." At the head of each palace hierarchy stood the wanax, the early Greek word for "king" or "ruler," written, according to Linear B spelling rules, as *wa-na-ka*. (The word's descendant, *anax*, meaning "lord" or "master," turns up five hundred years later in Homeric Greek.) The wanax was the administrative leader of each Mycenaean kingdom, overseeing domestic economics and foreign trade, military preparedness, ritual observance, law, and, in an inevitability that seems a hallmark of every human civilization, taxation. "His status," Shelmerdine explains, "is reflected in his superior land holdings." From one Pylos tablet, for instance, we know that "his temenos, or plot of land, is three times as big as those of other officials listed there."

The picture of the wanax that emerges most clearly is that

of an economic head of state. As Shelmerdine points out, Linear B records take pains to identify certain craftsmen—"a potter, a fuller and an armourer at Pylos, a textile worker at Thebes, and purple dye workers at Knossos"—by the designation *wanakteros* (spelled *wa-na-ka-te-ro),* an adjective meaning "royal." These were, in other words, the handpicked craftsmen to the king; their skill had elevated them to positions like those of the present-day British firms that by royal warrant may call themselves "Stationer/Robe Maker/Wheelwright . . . to the Queen."

Just below the wanax was the lawagetas (spelled *ra-wa-ke-ta),* or feudal landowner. The lawagetas—there was one each at Knossos and Pylos—seemed to be in charge of certain groups of subordinates, who helped with the day-to-day running of the kingdom. These subordinates included some military personnel, like rowers, as well as smallholders, who appear to have presented the lawagetas with a share of their agricultural yield in exchange for being granted land to work.

Below the lawagetas were still other officials, including hekwetai (*e-qe-ta*; literally "followers"), high-level representatives of each palace who appear to have had military responsibilities; "collectors," who seem to have been comptrollers, responsible for palace commodities including livestock; and the scribes themselves, whose literacy skills were essential for palace record-keeping. At the regional level, officials included damokoros, or provincial governors; at the local level they included mayors and vice mayors; landowners known as telestai *(te-re-ta);* and "fig-overseers."

MYCENAEANS PLIED a range of trades. Many tablets reveal the names of occupations—they appear, for instance, on lists of men assigned to military details; on inventories of raw materials issued to metalsmiths; and on accountings of rations dispensed to indentured servants and their dependent children—and from these lists it is possible to tell quite a lot about who did what in the Mycenaean world.

Most workers listed on the tablets were men, but from those on which lists of women's names appear, it is clear that certain occupations, like textile work, were traditionally reserved for them. Women spun sheep's fleece into woolen yarn and flax into linen and wove it into cloth on looms; men took the cloth, fulled it (a process, analogous to felting, that strengthens and stabilizes the fabric), and dyed it. The tablets also mention tanners, and leatherworkers of both sexes: Men fashioned the leather into harnesses, while women stitched it into shoes and bags. There were also men and women in religious life, the priests and priestesses.

Men were involved in the making of war (soldiers, rowers, and archers) and in the manufacture and upkeep of the instruments of war (swordmakers and bowmakers, chariotmakers and chariot-wheel repairmen). They worked as goldsmiths and perfumers, a major enterprise in the Mycenaean world. There were woodcutters, carpenters, shipbuilders, and netmakers; fire kindlers and bath attendants; heralds, hunters, herdsmen, and beekeepers.

There were also slaves. One tablet, from Knossos, records the acquisition of a slave. Others, from various sites, list rations of grain (wheat or barley), figs, and bedding disbursed to female slaves and their children. As Shelmerdine writes, "The tablets reinforce the view" that Mycenaean society comprised "two kinds of people: the social/political/economic elite, and those who do their work and supply their particular needs. The texts thus present an array of different craftsmen and herdsmen, who must have occupied the middle levels of society, as well as fully dependent workers housed and fed by the palace." Some of these dependent workers, both men and women, are described on the tablets as doelos (*do-e-ro*) and doela (*do-e-ra*), respectively; both terms are akin to the Classical Greek word *doulos*, "slave."

At the palaces, resident groups of women were assigned to perform specific tasks, including weaving: Cloth was a valuable commodity in overseas trade. "Slave status is suggested for these women because they are fully supported with rations by the palace, appear in groups rather than as individuals, and are not named," Shelmerdine writes. Many of them were foreigners, imported to work in the Mycenaean palaces, as the tablets make clear: At Pylos, one such group is described as "captives"; others, Shelmerdine says, "are identified by [non-Greek] ethnic adjectives: Milesians, Knidians, Lemnians, Lydians and so on." Outside the palaces, certain craftsmen, like bronzesmiths, appeared to have had male slaves assigned to aid their labors.

Because many Linear B tablets contain personal names, we know quite a bit about early Greek naming practices. Some men's names are descriptive, and evocatively so, with English equivalents like "Gladly Welcome," "Head of the Community," "Born on the Third Day of the Month," "Snub-Nosed," and, less flatteringly, "Coward." Other names are followed by descriptive epithets, often heroic in nature: ". . . Who Commands the People," ". . . Who Remembers His Work," ". . . Who Overcomes Men," ". . . Who Kills in Battle," ". . . Who Watch Fire."

Still other people had names that while "highly expressive," as J. L. García Ramón writes, were "anything but heroic." Among them are "Goat-Head," "Mouse-Head," "Having the Bottom Bare," and "Devourer of Excrements." Such names, perhaps unsurprisingly, appear to have been bestowed upon foreigners and slaves.

THE TABLETS ARE economic documents above all, and in the premonetary society of Mycenae, economics was rooted in the amassing, enumeration, and exchange of goods: livestock, agricultural produce, and man-made wares. Many tablets did indeed, as Kober suggested, keep track of cattle. Others are quite literally devoted to counting sheep, and there were an awful lot of sheep to count: A group of eight hundred tablets from Knossos alone inventories nearly a hundred thousand of them. Mycenaean scribes also kept tabs on pigs, cattle, goats, oxen, and other animals, both in flocks and as the stuff of state banquets. Charmingly, the tablets sometimes record the names of individual oxen—names with English equivalents like "Changeful of Hue," "Noisily Prattling," "Dapple," "Winey," and "Blondie." Still other tablets count the kingdom's horses: the now-famous tablet with the Linear B word "po-lo" (*pōlos*) was one of them.

Agriculture was as important as livestock to the Mycenaean economy. Besides wheat, barley, and figs, important agricultural products included olives, olive oil, and pistachio nuts; wine, cheese, and honey; spices like saffron and coriander; and flax. Many commodities were considered so vital that special offices in the palaces existed just to keep track of them: At Knossos, an office in the east wing of the Palace of Minos housed records having to do with honey and aromatics; another, in the west wing, handled records pertaining to sheep.

The production of wheat was an especially vast enterprise, as the tablets attest. On Crete, as John Chadwick writes in *The Mycenaean World*:

> The most extraordinary figure for wheat is for the area called *Dawos,* which we have good reason to think was in the fertile plain of the Messará in the south of the island. Here the tablet is broken so that the numeral is incomplete, but it unquestionably began with 10,000 units. Even assuming that no further figures followed, this would amount to some 775 tons. . . .

On the mainland, he writes:

> The absence of any record of the grain harvest at Pylos is doubtless due to the time of the year at which the destruction of the archives occurred. . . . But we can infer something about the scale of production from the rations issues to the slave-women. . . . A broken tablet . . . is probably a total of the rations issued each month to these women; it gives a figure of 192.7 units of wheat, or around 14 tons. This implies the need for an annual production of about 170 tons for this purpose alone.

Though the Mycenaeans had no money as we know it, they most assuredly paid taxes, and many Linear B tablets turn out to be tax records. The central palaces exacted payment from their constituent districts in the form of raw materials and other articles of value, including oil, olives, grain, honey, spices, horn, wood, animal hides, and cyperus, an aromatic grass. The tablets also show that members of certain professions, including bronzesmiths, had tax-exempt status: They were relieved of the obligation to make payments in kind, though it can be assumed that their contributions were taken out in the form of labor.

THE MAKING OF useful and beautiful things—from chariots, wheels, and weapons to furniture, vessels, textiles, and perfume—was a thriving enterprise in the Mycenaean world, and the results are copiously recorded in its archives.

Four metals are mentioned on the tablets. Foremost was bronze, "used for a variety of purposes and no doubt . . . the most important metal for everyday life in the upper classes," as the scholars Alberto Bernabé and Eugenio R. Luján write. It was used, among other things, to make vessels, braziers, certain chariot components, and weaponry, including the heads of spears and javelins. (Such spearheads, the two scholars point out, are mentioned in Homer, as in these lines from book 4 of the *Iliad*: "He held a spear of eleven cubits; / at the front, the bronze spear-point blazed.") Gold was used to decorate furniture, silver to decorate chariot wheels. Lead was also used: Recall the lead-lined cist full of tablets discovered in Emperor Nero's day.

From wood, furniture was made, and it is inventoried extensively in the Linear B records. Wooden furniture included stools, chairs, beds, and the nine-legged tables, characteristic of Mycenaean carpentry, that were often inlaid with ivory (the Mycenaean word for which was *elephas*, spelled *e-re-pa)*, gold, lapis, or other precious materials. Chairs, recorded as being made of ebony, could also be lavishly ornamented. They often had matching footstools, itemized on the tablets with the logogram [illegible].

Chariots and their wheels were also made of wood. Drawn by two horses, a Mycenaean chariot could accommodate two men, the driver and a warrior. Their building and upkeep were so essential to the well-being of the kingdoms that the Linear B archives contain detailed chariot construction and maintenance records. (Alice Kober's seminal paper of 1945, "Evidence of Inflection in the 'Chariot' Tablets from Knossos," concerned a set of these tablets.) From such tablets, Bernabé and Luján write:

> We are well informed about the chariot's constituent elements. . . . The frame of the case was made of wood and covered by leather at the front and at the sides. The chariot floor . . . most probably consisted of flexible leather straps. In order to make access easier the chariots were provided with footboards. . . . Wheels turned on an axle. On [one Pylos tablet] thirty-two bronze axles . . . are recorded. . . .

Another tablet from Pylos, they write, "is a delivery record concerning the fabrication of chariots and wheels," and makes note of wooden axles. It reads, "Thus the wood-cutters give to the wheeler's workshop 50 new branches and 50 axles."

The production of woven textiles, a major industry, is also well documented. Woolens are sometimes described as being white or gray, natural colors for undyed fleece. But the tablets show that the Mycenaeans also produced colored textiles, dyed in hues of purple and red, with dyes made from natural ingredients like minerals and plants. They also describe the production of various kinds of cloth. One type, pharweha (*pa-we-a*), has a Homeric counterpart in *pharos*: In the *Odyssey*, the word denotes the cloth the faithful Penelope weaves—and rips out nightly to thwart a stream of gentleman callers—while she waits for her husband, Odysseus, to return from the Trojan War.

Potters and metalsmiths made storage containers of all sorts, and the archives list not only the vessels themselves but also what was kept in them. "Remarkable at Pylos is the amount of ground floor space given over to storage," Cynthia Shelmerdine writes. "Hundreds of drinking and eating vessels stood on shelves in the pantries; storerooms housed olive oil, some of it perfumed, as tablets found with the storage jars make clear. More oil tablets fell from above when the building burnt down *ca* 1200 BC, so another such storeroom must have stood on the upper storey. Other objects that fell from above include jewellery, ivory inlays from furniture, and tablets dealing with linen textiles. These finds suggest that the upstairs too was a mix of private quarters, storage and business areas."

On Crete and in mainland Greece, archaeologists have unearthed Mycenaean "stirrup-jars" (traditional clay vases with handles and spouts, used to hold oil and wine); the jars were often painted with Linear B words. The texts they bear are utilitarian—they tend to document the production, transport, and delivery of the jar's contents—functioning much as inventory, shipping, and tracking labels on modern packages do.

The Mycenaeans were master perfumers. "The production of perfumes," Bernabé and Luján write, "was one of the most important industrial activities in Mycenaean times," and many storage vessels contained perfumed oil. The steps involved in making perfume, or "ointment," as the tablets often called it, are well documented: First, the ointment-maker infused wine with spices like cumin, coriander, fennel, sesame, or sage—or with herbs and flowers like rush, rose, and perhaps iris—to extract their fragrance. Other ingredients, like fruit, honey, and possibly lanolin, might be included.

Next, he added a thickener like natural gum or resin to a quantity of olive oil. The wine and herbs were mixed with the oil and the mixture was concentrated by boiling. Finally, a coloring agent like henna might be added to the finished perfume. These fragrant products appear to have been used for personal adornment as well as in religious ritual, offered frequently as gifts to the gods. (Mycenaean cloth and perfumed oil were also traded overseas for precious metals like gold and silver.)

THE TABLETS ALSO document the threat of war. Though there was no Talos, the bronze giant said to have policed the Cretan coast, it is clear from the tablets that all the kingdoms took great pains to protect against attack. This imperative appears to have arrived with the Greeks. As John Chadwick writes in *The Mycenaean World*: "Minoan society in Crete seems to have been relatively peaceful; military scenes are not common in art. . . . No Minoan town seems to have been fortified. But with the coming of the Greeks to Crete in the . . . fifteenth century, a change comes over the pacific face of society. . . . Greek rule in Crete is distinguished by this warlike aspect."

Many Linear B records deal with military preparedness: Some tablets itemize the horses, chariots, weapons, and other equipment issued to soldiers. This included suits of armor, comprising a helmet with earflaps (identified by the logogram) and a leather corselet with shoulder pieces (), along with arrows, spears, and swords, some inlaid with silver or gold.

Other tablets list the names of men, including archers and rowers, assigned to military duty on land or sea. A tablet from Pylos records men's names under the heading, "Thus the watchers are guarding the coastal regions." Another lists eight hundred rowers assigned to patrol various points along the shoreline. Similar conscription records are found at Knossos.

RELIGION CAN BE found between the lines. Much has been gleaned obliquely from certain inventories on the tablets, including lists of gifts to the gods and supplies for religious feasts. "The Linear B documents concern the economic administration of the palace in its various aspects," the scholar Stefan Hiller writes. "Therefore, there are no religious texts in the strict sense of the word—no prayers, hymn, manuals of religious instruction. All that we can use are the records of economic transactions. . . . In addition the records that list palace personnel or provide for their subsistence sometimes mention titles of religious dignitaries."

The tablets capture a theology in transition. On the one hand, Hiller writes, they offer "striking proof of a high degree of continuity between Mycenaean and Classical Greek religion": Gods' names listed there include some of the most renowned figures in the Olympian pantheon, like Dionysus (long thought by scholars not to have appeared till the first millennium B.C.), Zeus, Poseidon, Hera, and Artemis.

But these names appear side by side with more curious ones, many of them pre-Greek, long forgotten by Classical times.

Among them are various female names—most likely those of local deities—beginning with the word *potnia,* "mistress": Mistress of Wild Beasts, Mistress of Horses, Mistress of Grain, Mistress of Asia, Mistress of the Labyrinth. The tablets also mention a few goddesses who were early female counterparts of male Olympians. They include Posidaeia, the opposite number of Poseidon, and Diwjā, that of Zeus. They, too, were gone by the Classical Age.

Some tablets contain lists of gifts offered to the gods. One, from Knossos, records twenty-two linen cloths presented to the Mistress of the Labyrinth. Others note gifts of manufactured items like gold vessels and perfume, as well as agricultural products like oil, olives, barley, spelt, figs, spices, wool, honey, and wine. Quantities of livestock, including sheep, goats, cattle, and pigs, were also sacrificed in the gods' honor: As recorded on one tablet from Pylos, Hiller writes, "3 bulls are sent by military contingents (troops) to the *di-wi-je-we e-re-u-te-re,* presumably the 'priest of Zeus.' "

Mycenaean state banquets served theological as well as political ends. "We know that from Homer onwards these banquets included a religious section, when the animals were slaughtered," Hiller writes. "That in Pylos state banquets were performed was concluded from archaeological evidence even before it was understood that several important tablets concerned this topic."

One tablet from Pylos documents the supplies required for such a banquet, possibly the initiation ceremony for the wanax. These include, Chadwick writes, "1,574 litres of barley, 14½ litres of cyperus, 115 litres of flour, 307 litres of olives, 19 litres of honey, 96 litres of figs, 1 ox, 26 rams, 6 ewes, 2 he-goats, 2 she-goats, 1 fattened pig, 6 sows [and] 585½ litres of wine," adding: "The barley alone would provide rations for 43 people for a month." That the entertainment at these banquets included music is known both from a mural in the Pylos palace depicting lyre players and from a tablet from Thebes, recording rations dispensed to various personnel, including two lūrastāe (*ru-ra-ta-e*), "lyre players."

Such meticulous accounts of banqueting supplies may have had a very particular function in Mycenaean religious life. "The book-keeping testifies to the practice of piety toward the gods," Stefan Hiller writes. He continues:

> There are strong reasons to believe that the primary motivation for the scrupulous monitoring of all major and minor expenditure for offerings and other religious activities was not only economic interest; it was much more the awareness that the communal welfare depended on the fulfillment of religious duties. Consequently it was the palace's most important obligation to secure the gods' benevolence through a firm control of all religious prescriptions. . . . The main purpose of writing offering lists may have been to make sure that all religious duties had been observed.

BUT FOR ALL THIS, the communal welfare of the Mycenaean state did not last forever. On Crete, catastrophe claimed the Palace of Knossos sometime between 1450 and 1400 B.C. On the mainland, Pylos, at least, held on until about 1200, when it, too, was destroyed. "What actually happened remains a tantalizing mystery," John Chadwick has written. "All we know is that the palace was looted and burnt. The absence of human remains suggests that no resistance took place there; . . . the archaeological picture suggests that the population was reduced to something like a tenth of its earlier numbers."

So ended the first flush of Greek civilization, and from then till the coming of the Greek alphabet centuries later, the art of writing was at best a dimly remembered dream. Before long the Mycenaean archives—describing a world of monarchs and slaves, gods and goddesses, spinners and weavers, men who made art and men who made war—had passed from readability into darkness, where they would languish for three thousand years.

That we have been able to admit this world to the annals of history owes to a long confluence of natural forces and forceful natures. Had the ancient palaces not burned to the ground; had Schliemann not dug at Mycenae; had Arthur Evans not been so very determined (and so very nearsighted); had Alice Kober not painstakingly scissored 180,000 index cards from odd scraps of paper; had Michael Ventris not been such a woeful boy, in deep need of intellectual distraction, we would know nothing of the written records of these early Greeks—the Bronze Age heroes of whom Homer would sing—unearthed, unlocked, and readable once more.

APPENDIX: THE SIGNS OF LINEAR B

𐀀 = “a”	𐀁 = “e”	𐀂 = “i”	𐀃 = “o”	𐀄 = “u”
𐀅 = “da”	𐀆 = “de”	𐀇 = “di”	𐀈 = “do”	𐀉 = “du”
𐀊 = “ja”	𐀋 = “je”		𐀍 = “jo”	𐀎 = “ju”
𐀏 = “ka”	𐀐 = “ke”	𐀑 = “ki”	𐀒 = “ko”	𐀓 = “ku”
𐀔 = “ma”	𐀕 = “me”	𐀖 = “mi”	𐀗 = “mo”	𐀘 = “mu”
𐀙 = “na”	𐀚 = “ne”	𐀛 = “ni”	𐀜 = “no”	𐀝 = “nu”
𐀞 = “pa”	𐀟 = “pe”	𐀠 = “pi”	𐀡 = “po”	𐀢 = “pu”
𐀣 = “qa”	𐀤 = “qe”	𐀥 = “qi”	𐀦 = “qo”	
𐀨 = “ra”	𐀩 = “re”	𐀪 = “ri”	𐀫 = “ro”	𐀬 = “ru”
𐀭 = “sa”	𐀮 = “se”	𐀯 = “si”	𐀰 = “so”	𐀱 = “su”
𐀲 = “ta”	𐀳 = “te”	𐀴 = “ti”	𐀵 = “to”	𐀶 = “tu”
𐀷 = “wa”	𐀸 = “we”	𐀹 = “we”	𐀺 = “wo”	
𐀼 = “za”	𐀽 = “ze”		𐀿 = “zo”	

𐁀 = “a”	𐁁 = “a”	𐁂 = “au”
𐁃 = “dwe”	𐁄 = “dwo”	𐁅 = “nwa”
𐁆 = “pu”	𐁇 = “pte”	
𐁈 = “ra”	𐁉 = “ra”	𐁊 = “ro”
𐁋 = “ta”	𐁌 = “twe”	𐁍 = “two”

𐁐, 𐁑, 𐁒, 𐁓, 𐁔, 𐁕, 𐁖, 𐁗, 𐁘, 𐁙, 𐁚, 𐁛, 𐁜, 𐁝 = *values still unknown.*

REFERENCES

"Adventure by Research Proves Ample Reward," *Brooklyn Eagle*, April 28, 1946.

Annual of the British School at Athens, no. 6, Session 1899–1900. London: Macmillan, [1901].

"Awards of University Scholarships." *New York Times*, Aug. 22, 1924, 16.

Barber, E. J. W. *Archaeological Decipherment: A Handbook*. Princeton: Princeton University Press, 1974.

Bennett, Emmett L., Jr. "The Minoan Linear Script from Pylos." Ph.D. dissertation, University of Cincinnati, 1947.

———. *The Pylos Tablets: A Preliminary Transcription*. Princeton: Princeton University Press, 1951.

Bernabé, Alberto, and Eugenio R. Luján. "Mycenaean Technology." In Duhoux and Morpurgo Davis (2008), 201–33.

Bliss, Charles Kasiel. *Semantography: A Non-Alphabetical Symbol Writing, Readable in All Languages; a Practical Tool for General International Communication, Especially in Science, Industry, Commerce, Traffic, etc., and for Semantical Education, Based on the Principles of Ideographic Writing and Chemical Symbolism*. 3 vols. Sydney: Institute for Semantography, 1949.

Brann, Eva. "In Memoriam: Alice E. Kober." Unpublished manuscript, Program in Aegean Scripts and Prehistory, University of Texas, Austin, 2005.

Candee, Marjorie Dent, ed. *Current Biography Yearbook*. New York: H. W. Wilson, 1958.

Candy, James S. *A Tapestry of Life: An Autobiography.* Braunton, UK: Merlin Books, 1984.

Casson, S., ed. *Essays in Aegean Archaeology: Presented to Sir Arthur Evans in Honour of His 75th Birthday.* Oxford: Clarendon Press, 1927.

Chadwick, John. *The Decipherment of Linear B.* Cambridge: Cambridge University Press, 1958; second edition, 1967.

———. "Greek Records in the Minoan Script." *Antiquity* 108 (1953), 196–206; includes "A Note on Decipherment Methods," by Michael Ventris.

———. *The Mycenaean World.* Cambridge: Cambridge University Press, 1976.

Christoforakis, J. M. *Knossos Visitor's Guide.* 3rd ed. Heraklion: n.d.

Conan Doyle, Sir Arthur. "The Adventure of the Dancing Men" (1903). In Conan Doyle (1960), vol. 2, 511–26.

———. *The Complete Sherlock Holmes.* With a preface by Christopher Morley. 2 vols. Garden City, NY: Doubleday, 1960.

Cottrell, Leonard. "Michael Ventris and His Achievement." *Antioch Review* 25:1 (1965), 13–40.

Coulmas, Florian. *The Blackwell Encyclopedia of Writing Systems.* Oxford: Blackwell, 1996.

Cowley, A. E. "A Note on Minoan Writing." in Casson (1927), 5–7.

Daniel, John Franklin. "Prolegomena to the Cypro-Minoan Script." *American Journal of Archaeology* 45:2 (1941), 249–82.

Dow, Sterling. "Minoan Writing." *American Journal of Archaeology* 58:2 (1954), 77–129.

Duhoux, Yves. "Mycenaean Anthology." In Duhoux and Morpurgo Davies (2008), 243–393.

Duhoux, Yves, and Anna Morpurgo Davies, eds. *A Companion to Linear B: Mycenaean Texts and Their World.* vols. 1–2. Louvain-la-Neuve: Peeters, 2008–2011.

"843 to Get Degrees at Hunter College." *New York Times*, June 14, 1928, 24.

Evans, A. J. "The Prehistoric Acropolis of Knossos." *ABSA* (1899–1900), 3–70.

Evans, Arthur J. "Further Discoveries of Cretan and Aegean Script: With Libyan and Proto-Egyptian Comparisons." *Journal of Hellenic Studies* 17 (1897), 327–95.

———. *Illyrian Letters: A Revised Selection of Correspondence from the Illyrian Provinces of Bosnia, Herzegovina, Montenegro, Albania, Dalmatia, Croatia and Slavonia Addressed to the "Manchester Guardian" during the Year 1877.* London: Longmans, Green, 1878.

———. "Pre-Phoenician Writing in Crete, and Its Bearings on the History of the Alphabet." *Man* 3 (1903), 50–55.

———. "Primitive Pictographs and a Prae-Phoenician Script, from Crete and the Peloponnese." *Journal of Hellenic Studies* 14 (1894), 270–372.

———. *Scripta Minoa: The Written Documents of Minoan Crete With Special Reference to the Archives of Knossos.* Vol. 1, Oxford: Clarendon Press, 1909; Vol. 2, edited by John L. Myres, Oxford: Clarendon Press, 1952.

———. *Through Bosnia and the Herzegóvina on Foot during the Insurrection, August and September 1875: With an Historical Review of Bosnia, and a Glimpse at the Croats, Slavonians, and the Ancient Republic of Ragusa*. London: Longmans, Green, 1876.

———. "Writing in Prehistoric Greece." *Journal of the Anthropological Institute of Great Britain and Ireland* 30 (1900), 91–93.

Evans, Arthur John. "On a Hoard of Coins Found at Oxford, with Some Remarks on the Coinage of the First Three Edwards." *Numismatic Chronicle* 11 (1871), 264–82.

Evans, Sir Arthur. *The Palace of Minos: A Comparative Account of the Successive Stages of the Early Cretan Civilization as Illustrated by the*

Discoveries at Knossos. 4 vols. in 6. Vol. 1 (1921); Vol. 2, in 2 parts (1928); Vol. 3 (1930); Vol. 4, in 2 parts (1935). London: Macmillan, 1921–35.

Evans, Joan. *Time and Chance: The Story of Arthur Evans and His Forebears.* London: Longmans, Green, 1943.

Fox, Margalit. "Linguistic Reanalysis and Oral Transmission." *Poetics: Journal of Empirical Research on Culture, the Media and the Arts* 13:3 (1984), 217–38.

———. *Talking Hands: What Sign Language Reveals About the Mind.* New York: Simon & Schuster, 2007.

Friedrich, Johannes. *Extinct Languages.* Translated by Frank Gaynor. New York: Philosophical Library, 1957.

García Ramón, J. L. "Mycenaean Onomastics." In Duhoux and Morpurgo Davies (2011), 213–51.

Garvin, Paul L. *On Linguistic Method: Selected Papers.* The Hague: Mouton, 1964.

Gere, Cathy. *Knossos and the Prophets of Modernism.* Chicago: University of Chicago Press, 2009.

Grote, George. *A History of Greece.* 12 vols. London: J. Murray, 1846–56.

"Guggenheim Fund Makes 132 Awards." *New York Times,* April 15, 1946, 16.

Haarmann, Harald. *Universalgeschichte der Schrift.* Frankfurt: Campus Verlag, 1990.

Hahn, E. Adelaide. "Alice E. Kober." Obituary note, *Language* 26:3 (1950), 442–43.

Hanff, Helene. *84, Charing Cross Road.* New York: Grossman, 1970.

Harden, D. B. *Sir Arthur Evans, 1951–1941: A Memoir.* Oxford: Ashmolean Museum, 1983.

Hiller, Stefan. "Mycenaean Religion and Cult." In Duhoux and Morpurgo Davies (2011), 169–211.

Homer. *The Odyssey.* Translated by Robert Fitzgerald. Garden City, NY: Anchor Books, 1963.

Horwitz, Sylvia L. *The Find of a Lifetime: Sir Arthur Evans and the Discovery of Knossos.* New York: Viking Press, 1981.

"J. F. Daniel 3d Dies; Archaeologist, 38." Obituary, *New York Times,* Dec. 19, 1948, 76.

Kahn, David. *The Codebreakers: The Story of Secret Writing.* New York: Macmillan, 1967.

Kempton, Murray. "When Constabulary Duty's to Be Done." *New York Newsday,* May 11, 1990.

Killen, J. T. "Mycenaean Economy." In Duhoux and Morpurgo Davies (2008), 159–200.

Kober, A. E. Review of Bedřich Hrozný, *Die Älteste Geschichte Vorderasiens und Indiens and Kretas und Vorgriechenlands Inschriften, Geschichte und Kultur—I, Ein Entzifferungsversuch. American Journal of Archaeology* 50:4 (1946), 493–95.

Kober, Alice E. "The 'Adze' Tablets from Knossos." *American Journal of Archaeology* 48:1 (1944), 64–75.

———. "The Cryptograms of Crete." *Classical Outlook* 22:8 (1945a), 77–78.

———. Curriculum vitae, n.d., c. 1947. Program in Aegaen Scripts and Prehistory, University of Texas, Austin.

———. "Evidence of Inflection in the 'Chariot' Tablets from Knossos." *American Journal of Archaeology* 49:2 (1945b), 143–51.

———. "Form Without Meaning." Text of lecture, Yale Linguistics Club, May 3, 1948 (=1948a).

———. "The Gender of Nouns Ending in -inthos." *American Journal of Philology* 63:3 (=1942a), 320–27.

———. "Inflection in Linear Class B: 1—Declension." *American Journal of Archaeology* 50:2 (=1946a), 268–76.

———. "The Language (or Languages) of the Minoan Scripts." Paper presented at meeting of the Classical Association of the Atlantic States (=1946b).

———. "The Minoan Scripts: Fact and Theory." *American Journal of Archaeology* 52:1 (1948b), 82–103.

———. Phi Beta Kappa lecture, no title, Hunter College, June 15, 1946 (=1946c).

———. Review of Bedřich Hrozný, *Les Inscriptions Crétoises: Essai de Déchiffrement* and Vladimir Georgiev, *Le Déchiffrement des Inscriptions Minoennes. Language* 26:2 (1950), 286–98.

———. "The Scripts of Pre-Hellenic Greece." *Classical Outlook* 21:7 (1944), 72–74.

———. "Some Comments on a Minoan Inscription (Linear Class B)." Paper presented at meeting of the Archaeological Institute of America, Dec. 31, 1941. Abstract published in *American Journal of Archaeology* 46:1 (=1942b), 124.

———. "Tiberius, Master Detective." *Classical Outlook* 22:4 (=1945c), 37.

———. " 'Total' in Minoan (Linear Class B)." *Archiv Orientálni* 17 (1949), 386–98.

Kober, Alice Elizabeth. "The Use of Color Terms in the Greek Poets, Including All the Poets from Homer to 146 B.C. Except the Epigrammatists." Ph.D. dissertation, Columbia University, 1932.

Ludwig, Emil. *Schliemann of Troy: The Story of a Goldseeker.* New York: G. P. Putnam's Sons, 1931.

McDonald, William A., and Carol G. Thomas. *Progress into the Past: The Rediscovery of Mycenaean Civilization.* 2nd ed. Bloomington: Indiana University Press, 1990.

MacGillivray, Joseph Alexander. *Minotaur: Sir Arthur Evans and the Archaeology of the Minoan Myth.* London: Pimlico, 2001.

Moorehead, Caroline. *Lost and Found: The 9,000 Treasures of Troy—Heinrich Schliemann and the Gold That Got Away.* New York: Viking, 1994.

Myres, J. L. "Arthur John Evans, 1951–1941." *Obituary Notices of Fellows of the Royal Society* 3:10 (Dec. 1941), 941–68.

———. "The Cretan Exploration Fund: An Abstract of the Preliminary Report of the First Season's Excavations." *Man* 1 (1901), 4–7.

Palaima, Thomas G. "Alice Elizabeth Kober." Unpublished manuscript, Program in Aegean Scripts and Prehistory, University of Texas, Austin (n.d.).

———. "Scribes, Scribal Hands and Palaeography." In Duhoux and Morpurgo Davies (2011), 33–136.

Palaima, Thomas G., Elizabeth I. Pope, and F. Kent Reilly III. *Unlocking the Secrets of Ancient Writing: The Parallel Lives of Michael Ventris and Linda Schele and the Decipherment of Mycenaean and Mayan Writing.* Exhibition catalogue, Nettie Lee Benson Latin American Collection, University of Texas. Austin: Program in Aegean Scripts and Prehistory, 2000.

Palaima, Thomas G., and Susan Trombley. "Archives Revive Interest in Forgotten Life." *Austin American-Statesman,* Oct. 27, 2003, A9.

Pope, Maurice. *The Story of Archaeological Decipherment: From Egyptian Hieroglyphs to Linear B.* New York: Charles Scribner's Sons, 1975.

"Prof. Alice Kober of Brooklyn Staff." Obituary, *New York Times,* May 17, 1950, 29.

Pyles, Thomas. *The Origins and Development of the English Language.* 2nd ed. New York: Harcourt Brace Jovanovich, 1971.

Robinson, Andrew. *Lost Languages: The Enigma of the World's Undeciphered Scripts.* London: Peter N. Nevraumont/BCA (=2002a).

———. *The Man Who Deciphered Linear B: The Story of Michael Ventris.* London: Thames & Hudson (=2002b).

———. *The Story of Writing* (London: Thames & Hudson, 1995).

Shelmerdine, Cynthia W. "Mycenaean Society." In Duhoux and Morpurgo Davies (2008), 115–58.

Singh, Simon. *The Code Book: The Science of Secrecy from Ancient Egypt to Quantum Cryptography.* New York: Anchor Books, 2000.

Smith, Prue. *The Morning Light: A South African Childhood Revalued.* Cape Town: David Philip, 2000.

Sundwall, Johannes. *Altkretische Urkundenstudien.* Åbo: Åbo Akademi, 1936.

———. *Knossisches in Pylos.* Åbo: Åbo Akademi, 1940.

———. "Minoische Rechnungsurkunden." *Societas Scientiarum Fennica, Commentationes Humanarum Litterarum* 4:4 (1932).

van Alfen, Peter G. "The Linear B Inscribed Vases." In Duhoux and Morpurgo Davies (2008), 235–42.

Ventris, M. G. F. "Introducing the Minoan Language," *American Journal of Archaeology* 44:4 (1940), 494–520.

Ventris, Michael. "Deciphering Europe's Earliest Scripts." Text of BBC Radio talk, first broadcast July 1, 1952. In Ventris (1988), 363–67.

———. "The Decipherment of the Mycenaean Script." *Proceedings of the Second International Congress of Classical Studies.* Copenhagen: Ejnar Munksgaard, 1958, 69–81.

———. "King Nestor's Four-Handled Cups: Greek Inventories in the Minoan Script." *Archaeology* 7:1 (Spring 1954), 15–21.

———. "A Note on Decipherment Methods." In Chadwick (1953).

———. *Work Notes on Minoan Language Research and Other Unedited Papers.* Edited by Anna Sacconi. Rome: Edizioni dell'Ateneo, 1988.

Ventris, Michael, and John Chadwick. *Documents in Mycenaean Greek: Three Hundred Selected Tablets from Knossos, Pylos and Mycenae with Commentary and Vocabulary.* Cambridge: Cambridge University Press, 1956.

———. "Evidence for Greek Dialect in the Mycenaean Archives." *Journal of Hellenic Studies,* 73 (1953), 84–103.

A Very English Genius. BBC documentary, originally broadcast 2002.

Wheelock, Frederic M. *Latin: An Introductory Course Based on Ancient Authors.* New York: Barnes & Noble Books, 1956.

Wilford, John Noble. "Greek Tablet May Shed Light on Early Bureaucratic Practices." *New York Times,* April 5, 2011, D3.

NOTES

Abbreviations

ABSA	*Annual of the British School at Athens*
AE	Sir Arthur Evans
AEK	Alice Elizabeth Kober
AJE	Arthur John Evans
CWB	Carl W. Blegen
ELB	Emmett L. Bennett Jr.
HAM	Henry Allen Moe
JC	John Chadwick
JFD	John Franklin Daniel
JLM	John Linton Myres
JS	Johannes Sundwall
LV	Lois Ventris
MV	Michael Ventris
PASP	Program in Aegean Scripts and Prehistory, University of Texas, Austin
WTMF	William T.M. Forbes

INTRODUCTION

xvi *likened to that of Rosalind Franklin*: Andrew Robinson, *Lost Languages: The Enigma of the World's Undeciphered Scripts* (London: Peter N. Nevraumont/BCA, 2002a), 16; Thomas G. Palaima and Susan Trombley, "Archives Revive Interest in Forgotten Life," *Austin American-Statesman*, Oct. 27, 2003, A9.

xvi *sitting night after night at her dining table*: Brann (2005), 4, speaks of AEK's doing this.

xvii *"IBM machines"*: AEK postcard to ELB, April 4, 1950, ELB Papers, PASP.

"I don't like the idea of getting paid": AEK to JFD, Feb. 18, 1948, AEK Papers, PASP.

xxviii *"In the words of Ventris"*: Robinson (2002a), 91.

"a feeling for the fitness of things": AEK to HAM, Sept. 8, 1947, AEK Papers, PASP.

xx *"There is no thread"*: Robinson (2002a), 95.

PROLOGUE: BURIED TREASURE

4 *On March 23, 1900*: See, e.g., Sir Arthur Evans, *The Palace of Minos: A Comparative Account of the Successive Stages of the Early Cretan Civilization as Illustrated by the Discoveries at Knossos*, vol. 4, part 2 (London: Macmillan, 1935), 668.

thirty local workmen: Joan Evans, *Time and Chance: The Story of Arthur Evans and His Forebears* (London: Longmans, Green, 1943), 329. The book's author was the younger half sister of AE.

on a knoll bright with anemones and iris: AE diary entry, March 19, 1894, quoted in J. Evans (1943), 312.

his workmen's spades turned up fragments: For accounts of the early finds at Knossos, see, e.g., J. Evans (1943), 330ff.; Horwitz (1981), 95ff.; and J. L. Myres, "The Cretan Exploration Fund: An Abstract of the Preliminary Report of the First Season's Excavations," *Man* 1 (1901), 4–7.

5 *the historic basis of the enduring myth of the labyrinth*: See, e.g., A. J. Evans, "The Prehistoric Acropolis of Knossos," *Annual of the British School at Athens (ABSA)*, no. 6, session 1899–1900 (London: Macmillan, [1901]), 33.

"such a find," Evans wrote: AE letter to Sir John Evans, November 1900, quoted in J. Evans (1943), 335.

5 *In his first season alone*: J. Evans (1943), 332–33; Myres (1901), 5.
On March 30: Evans (1899–1900), 18.
On April 5: Ibid.

6 *more than a thousand tablets*: Myres, "The Cretan Exploration Fund," 5.

7 *a special font, in two different sizes*: Evans (1899–1900), 58, note 2; Horwitz (1981), 131.
"Of all the decipherments of history": David Kahn, *The Codebreakers: The Story of Secret Writing* (New York: Macmillan, 1967), 917.

CHAPTER ONE: THE RECORD-KEEPERS

13 *the most visible Bronze Age ruins there could be dated*: Horwitz (1981), 63.
the distinguished archaeologist Flinders Petrie: See, e.g., Joseph Alexander MacGillivray, *Minotaur: Sir Arthur Evans and the Archaeology of the Minoan Myth* (London: Pimlico, 2001), 78ff.

14 *Schliemann dug fruitlessly for several years*: Horwitz (1981), 61.
the authenticity of some of his finds: For a précis of the doubts cast on Schliemann's work, see, e.g., MacGillivray (2001), 57ff.

15 *"the later Greeks understandably concluded"*: John Chadwick, *The Mycenaean World* (Cambridge: Cambridge University Press, 1976), 55.
Digging down into the circle: MacGillivray (2001), 60ff.

16 *"It seemed incredible that [such] a civilisation"*: Arthur J. Evans, "Pre-Phoenician Writing in Crete, and Its Bearings on the History of the Alphabet," *Man* 3 (1903), 52.
Perhaps the Mycenaeans had written on perishable materials: See, e.g., Evans (1909), 3.
hints that they had written on sturdier stuff: A list of Mycenaean objects known as of 1894 to contain writing-like sym-

bols, including seal-stones, pottery, engraved gems, inscribed building blocks, and clay pendants, appears in Evans (1894), 346.

16 *unearthed a clay amphora*: Ibid., 273. This and the vase-handle discovery are credited to Tsountas in Evans (1903).

a stone vase whose handle was engraved: Arthur J. Evans, "Primitive Pictographs and a Prae-Phoenician Script, from Crete and the Peloponnese," *Journal of Hellenic Studies* 14 (1894), 273.

the remains of a Bronze Age wall: MacGillivray (2001), 93.

In the early 1880s: Ibid., 95.

dismissed as "masons' marks": Evans (1894), 281.

had a symbol in common: Ibid., 282–83.

17 *His father, Sir John Evans*: John Evans was knighted in 1892. J. Evans (1943), 302.

Evans the Great: Horwitz (1981), 6.

"helped to lay the foundations": Ibid., 9.

18 *"a bit of a dunce"*: J. Evans (1943), 93; also quoted in Horwitz (1981), 17.

On New Year's Day 1858: J. Evans (1943), 93.

Arthur's much younger half sister: Joan Evans (1893–1977), a noted antiquarian and art historian, was the daughter of John Evans by his third wife, Maria Millington Lathbury, born when her father was about seventy. Joan Evans was made a Dame of the British Empire in 1976.

"John Evans wrote in his wife's diary": J. Evans (1943), 94.

John Evans married a cousin: Ibid., 104.

Arthur won prizes in natural history: J. L. Myres, "Arthur John Evans, 1951–1941," *Obituary Notices of Fellows of the Royal Society*, 3:10 (Dec. 1941), 941; J. Evans (1943), 145.

graduating with first-class honors in 1874: D. B. Harden, *Sir Arthur Evans, 1951–1941: A Memoir* (Oxford: Ashmolean Museum, 1983), 10.

18 *he published his first scholarly article*: Arthur John Evans, "On a Hoard of Coins Found at Oxford, with Some Remarks on the Coinage of the First Three Edwards," *Numismatic Chronicle* 11 (1871), 264–82.

"Little Evans, son of John Evans the Great": Horwitz (1981), 6.

19 *He traversed the wild countryside*: See, e.g., Ibid., 40.

the first of his two books on the Balkans: Arthur J. Evans, *Through Bosnia and the Herzegóvina on Foot during the Insurrection, August and September 1875: With an Historical Review of Bosnia, and a Glimpse at the Croats, Slavonians, and the Ancient Republic of Ragusa* (London: Longmans, Green, 1876).

"Mind where you travel!": Quoted in J. Evans (1943), 194.

In September 1878, Evans married: Ibid., 214.

now best remembered for his fiercely held views: See, e.g., MacGillivray (2001), 52ff.

20 *Evans had horrified his father*: J. Evans (1941), 216.

after seven weeks in a local jail: J. Evans (1943), 258.

In 1884, Evans was appointed keeper: Harden (1983), 9.

A diminutive man of barely five feet: Horwitz (1981), 1, puts Evans's height at five foot two; MacGillivray (2001), 34, writes that he "never grew much beyond four feet."

he nonetheless bristled: J. Evans (1943), 202.

"I don't choose to be told": Evans (1876), 312; also quoted in J. Evans (1943), 202.

"But . . . it is easy to see how valuable": Evans (1876), 312; also quoted in J. Evans (1943), 202.

"His short sight": J. Evans (1943), 144.

21 *"like a jackdaw down a marrow bone"*: Quoted in Horwitz (1981), 22.

22 *the five months he and Margaret spent*: Horwitz (1981), 64.

the widely accepted view of Greek history: This position was advanced by the historian George Grote in his seminal work,

A History of Greece, published in twelve volumes between 1846 and 1856. For a discussion of Grote's influence, see, e.g., MacGillivray (2001), 57ff.

22 *and with it, recorded history*: Of the world's roughly six thousand languages, only a minority have writing systems, and many cultures from antiquity to the present day have relied entirely on oral tradition as a means of transmitting their own histories. However, orally transmitted texts almost inevitably undergo change—often considerable change—over time, through constant retelling. See, e.g., Margalit Fox, "Linguistic Reanalysis and Oral Transmission," *Poetics: Journal of Empirical Research on Culture, the Media and the Arts* 13:3 (1984), 217–38.

"at a comparatively low level of civilization": Chadwick (1976), 180.

"is a network of well-organized kingdoms": Ibid.

"When Homer describes a letter": Ibid., 182.

23 *his father and two colleagues had unearthed Stone Age implements*: See, e.g., Horwitz (1981), 7ff.; MacGillivray (2001) 30ff.

"that human beings had lived on this earth": MacGillivray (2001), 31.

at a Roman site in Trier: Horwitz (1981), 28.

"Such a conclusion": Arthur J. Evans, *Scripta Minoa: The Written Documents of Minoan Crete with Special Reference to the Archives of Knossos* (Oxford: Clarendon Press, 1909), 3.

24 *"The discoveries of Schliemann"*: Evans (1903), 51.

Evans had bought sixty acres: Horwitz (1981), 128.

"from the ancient name of the heath below": Ibid., 76.

In 1890, Margaret had been diagnosed: MacGillivray (2001), 82.

In March 1893, Margaret Evans died: Horwitz (1981), 77.

"For the rest of his life he wrote on black-edged paper": Ibid.

25 *completed the next year, and Evans moved into it alone*: J. Evans (1943), 306.

red or green jasper, carnelian, or amethyst: MacGillivray (2001), 74.

"a series of remarkable symbols": Evans (1894), 274.

"not a mere copy of Egyptian forms": Ibid., 371.

Schliemann's bead gems: Ibid., 272.

26 *"To Crete," Evans wrote*: Evans (1909), 10.

Crete's earliest known inhabitants: Evans (1894), 275.

"It was clearly recognized by the Greeks themselves": Ibid., 354.

Evans paid his first visit to Crete: J. Evans (1943), 310.

"In the evening some excitement": Quoted in Ibid., 311.

27 *"It is impossible to believe"*: Evans (1894), 300.

28 *"that the great days of the island"*: Evans (1909), 10.

"a clue to the existence of a system": Quoted in Horwitz (1981), 81. Evans made the announcement in November 1893, at a meeting of the Hellenic Society in London. See also Evans (1909), 9ff.

"an elaborate system of writing did exist": Evans (1894), 274.

"linear and quasi-alphabetic": Ibid.

"a kind of linear shorthand": Ibid., 367.

"Of this linear system too": Ibid., 363.

29 *"One of the great islands of the world"*: Homer, *The Odyssey*, Book 19, lines 172ff., translated by Robert Fitzgerald (Garden City, NY: Anchor Books, 1963), 359.

tou Tseleve he Kephala: MacGillivray (2001), 92.

The reasoning, which Kalokairinos accepted: Ibid., 93.

In the early 1880s, William James Stillman: Ibid., 95ff.

30 *Schliemann, too, had his eye on Kephala*: Caroline Moorehead, *Lost and Found: The 9,000 Treasures of Troy—Heinrich Schliemann and the Gold That Got Away* (New York: Viking, 1994), 213ff.

30 *"Nor can I pretend to be sorry"*: Sir Arthur Evans, introduction to Emil Ludwig, *Schliemann of Troy: The Story of a Goldseeker* (New York: G. P. Putnam's Sons, 1931), 19, quoted in Horwitz (1981), 87.

"seemed to belong to an advanced system of writing": Evans (1909), 17.

"On the hill of Kephala": Ibid.

In 1894, after much negotiation: Myres (1941), 947.

"native Mahometans": Evans (1899–1900), 5.

for 235 British pounds: J. Evans (1943), 319.

31 *"striking corroboration"*: Arthur J. Evans, "Further Discoveries of Cretan and Aegean Script: With Libyan and Proto-Egyptian Comparisons," *Journal of Hellenic Studies* 17 (1897), 393.

"long before our first records": Evans (1897), 393.

the last of the Turkish forces left the island in late 1899: J. Evans (1943), 326.

"after encountering every kind of obstacle": Evans (1909), 17.

for 675 pounds: J. Evans (1943), 321.

he equipped himself with: Ibid., 329; MacGillivray (2001), 166–67.

a fleet of iron wheelbarrows: Evans (1899–1900), 68.

he set about disinfecting and whitewashing: J. Evans (1943), 329.

with the Union Jack flying: MacGillivray (2001), 175.

about 6100 B.C.: Horwitz (1981), 96. The date was ascertained by later archaeologists, using carbon-14 dating.

32 *It was rebuilt and partly reoccupied*: Evans (1909), 53.

a building larger than Buckingham Palace: Horwitz (1981), 232.

spread over six acres: Sir Arthur Evans, *The Palace of Minos: A Comparative Account of the Successive Stages of the Early Cretan Civilization as Illustrated by the Discoveries at Knossos* (London: Macmillan), vol. 1 (1921), v.

32 *a small Egyptian statue, carved of diorite*: Evans (1899–1900), 27.
The palace comprised hundreds of rooms: J. M. Christoforakis, *Knossos Visitor's Guide*, 3rd ed. (Heraklion: n.d.), 27.
the 1900 season, which lasted nine weeks: Evans (1899–1900), 69.
30 workmen had grown to about 180: Ibid.

33 *Evans employed both Christian and Muslim workers*: Ibid.
In the course of the season, Evans's workers unearthed: See, e.g., Myres (1901), 5; Evans (1899–1900), passim.
So delighted was Sir John: J. Evans (1941), 333.
"almost thrown into the shade": Myres (1901), 6.
"a discovery which carries back": Ibid.
"a kind of baked clay bar": Quoted in J. Evans (1943), 330–31. A similar account appears in Evans (1899–1900), 18.

34 *"the dramatic fulfillment"*: Evans (1909), vi.
On April 5: Evans (1899–1900), 18.
small bronze hinges: Ibid., 29.
"just struck the largest deposit yet": Quoted in J. Evans (1943), 334.
between two and seven inches long: Arthur J. Evans, "Writing in Prehistoric Greece," *Journal of the Anthropological Institute of Great Britain and Ireland* 30 (1900), 92.
sometimes fashioned around armatures of straw: Chadwick (1976), 18.
One very large rectangular tablet: Evans (1909), 48.

36 *in use from about 2000 to 1650 B.C.*: John Chadwick, *The Decipherment of Linear B*, 2nd ed. (Cambridge: Cambridge University Press, 1967), 12.
Evans came across only a single cache: Evans (1899–1900), 59.
"a new system of linear writing": Evans (1900), 91.
"style of writing fundamentally different": Ibid., 92.
used from about 1750 to 1450 B.C.: Chadwick (1976), 13.

37 *"Evidently the tablets were supplied"*: Sir Arthur Evans, *The Palace of Minos*, vol. 4 (1935), 695.

38 *more than two thousand would be found there*: Chadwick (1976), 15.

"the work of practised scribes": AE letter to Sir John Evans, April 15, 1900. Quoted in J. Evans (1943), 333.

fingerprints and even occasional doodles: Chadwick (1976), 20, 25.

39 *"We have here locked up for us"*: Myres (1901), 6.

"The problems attaching to the decipherment": Ibid.

CHAPTER TWO: THE VANISHED KEY

42 *By some estimates, only about 15 percent*: Harald Haarman, *Universalgeschichte der Schrift* (Frankfurt: Campus Verlag, 1990), 18ff.

44 *"Decipherments are by far the most glamorous"*: Maurice Pope, *The Story of Archaeological Decipherment: From Egyptian Hieroglyphs to Linear B* (New York: Charles Scribner's Sons, 1975), 9.

Diagrammed, they make a tidy four-cell table: The table is after E. J. W. Barber, *Archaeological Decipherment: A Handbook* (Princeton: Princeton University Press, 1974), 13. This typology of decipherment was first put forth by Johannes Friedrich in, e.g., *Extinct Languages* (New York: Philosophical Library, 1957), translated by Frank Gaynor.

45 *a Polynesian language still spoken on the island*: www.ethnologue.com.

46 *Weighing three-quarters of a ton*: Simon Singh, *The Code Book: The Science of Secrecy from Ancient Egypt to Quantum Cryptography* (New York: Anchor Books, 2000), 205–6.

47 *"the benefits that the Pharaoh Ptolemy had bestowed"*: Singh (2000), 207.

47 *the demotic script had been introduced*: Andrew Robinson, *The Story of Writing* (London: Thames & Hudson, 1995), 16.

48 *Born in 1773*: Singh (2000), 207.

"Young was able to read fluently": Ibid., 207–8.

49 *"it would enable [him] to discover"*: Ibid., 209.

the arrangement of the symbols in a cartouche was rarely fixed: Ibid., 210.

50 *"Although he did not know it at the time"*: Ibid.

Here are the actual sound-values: The chart is adapted from Ibid., 209.

51 *"Latin, Greek, Hebrew, Ethiopic"*: Ibid., 213.

"that he used it to record entries in his journal": Ibid., 215.

52 *then the cartouche so far would read*: After Ibid.

stood for the consonant cluster "ms": Robinson (1995), 33.

53 *As Ventris's biographer Andrew Robinson points out*: Ibid., 101.

56 *Rongorongo contains hundreds of logograms*: Andrew Robinson, personal communication.

Arthur Conan Doyle's "Dancing Men" cipher: Robinson (2002) also invokes this cipher in a discussion of archaeological decipherment.

58 *the Rotokas alphabet of the Solomon Islands*: Robinson (1995), 169; www.omniglot.com.

the thirty-three Cyrillic letters used in Russian: Coulmas (1996), 109.

the more than seventy characters of the Khmer alphabet: Robinson (1995), 169.

It will take our alien years of minute comparison: The three capital *O*'s are set, respectively, in the typefaces Edwardian Script, Matisse, and Jokerman. The next three letters are a capital *O* and *Q*, both in French Script, and a capital *C* in Edwardian Script. Andrew Robinson makes a similar point about the challenge of identifying variant letter-forms in *The Man*

Who Deciphered Linear B: The Story of Michael Ventris (London: Thames & Hudson, 2002b), 64–65.

59 *"The individual signs of Linear B"*: Kahn (1967), 919. This passage is also quoted in Singh (1999), 220–21.

years agonizing over the symbol [Linear B sign]: See, e.g., AEK to JFD, May 4, 1942, AEK Papers, PASP; Alice E. Kober, "The 'Adze' Tablets from Knossos," *American Journal of Archaeology* 48:1 (1944), 65, note 2.

60 *There were perhaps seventy scribes at Knossos*: Chadwick (1976), 24.

it was not completed until 1951: Emmett L. Bennett Jr., *The Pylos Tablets: A Preliminary Transcription* (Princeton: Princeton University Press, 1951).

61 *The Germans call this style* Schlangenschrift: Thomas G. Palaima, personal communication.

62 *"The Adventure of the Dancing Men"*: Sir Arthur Conan Doyle, "The Adventure of the Dancing Men," in Sir Arthur Conan Doyle, *The Complete Sherlock Holmes* (Garden City, NY: Doubleday, 1960), with a preface by Christopher Morley, vol. 2, 511–26.

64 *small vertical tick marks*: This inscription is also reproduced in Robinson (1995), 111.

"documents of 'lime-bark' ": Evans (1935), 673; see also Evans (1909), 108ff.

65 *"The brown, half-burnt tablets"*: Evans (1935), 673.

Nero was reported to have ordered the documents translated: Evans (1909), 109.

CHAPTER THREE: LOVE AMONG THE RUINS

67 *the precious unbaked records reduced to mud*: See, e.g., Evans (1909), 43.

"In this way fire": Evans (1899–1900), 56.

68 *In both scripts, text was written from left to right*: Evans (1935), 684.

69 *"The conclusion has been drawn"*: Ibid., 711.

filed neatly away by subject: Ibid., 694.

what appeared to Evans to be census data: Ibid., 694, 708.

70 *Evans was able to work out the numerical system*: Ibid., 691ff.

71 *These often appeared next to numbers*: Evans (1921), 46–47.

the "Armoury Deposit": Evans (1935), 832.

more than eight thousand arrows inside: Evans (1909), 44.

depicting the trellises on which grapes were grown: Chadwick (1976), 124.

72 *male and female animals*: Evans (1935), 723, 801.

the elegant answer to this little puzzle: Chadwick (1967), 45.

logograms denoting vessels: Evans (1935), 727.

a tablet inscribed with pictures of humble pots like these: Chadwick (1967), 81ff.

David Kahn described so evocatively: Kahn (1967), 919.

73 *By Evans's initial count*: Evans (1903), 53.

"The number of signs between word boundaries": Barber (1974), 94–95.

modern Japanese writing: Coulmas (1996), 239ff.

Evans himself suspected as much: In a monograph on the early Cretan scripts (1903, 53), he wrote, "The characters seem to have had a syllabic value."

74 *"No effort will be spared"*: Evans (1899–1900), 59, note 2.

only about two dozen pages: Evans (1909), 28–54.

a feat he wisely deemed impossible: Ibid., v.

Although Evans promised additional volumes: Ibid., x.

the Knossos tablets remained locked away: Horowitz (1981), 159.

75 *Not only did he decline*: Evans did, apparently, share particular tablets with certain trusted individual scholars. See, e.g.,

A. E. Cowley, "A Note on Minoan Writing," in S. Casson, ed., *Essays in Aegean Archaeology: Presented to Sir Arthur Evans in Honour of His 75th Birthday* (Oxford: Clarendon Press, 1927), 5–7.

75 *he would publish reproductions of fewer than two hundred*: Chadwick (1967), 18.

an act that brought down the wrath of Evans: Ibid.; Evans (1935), 681, note 1.

He remained keeper of the Ashmolean till 1908: Harden (1983), 14.

serving more or less simultaneously as president: Ibid., 20.

the local Boy Scout troop: Horwitz (1981), 167–68.

he took in two wards: Ibid.

digging on Crete became impossible for the duration: Ibid., 182.

76 *taking an active hand in the negotiations*: Ibid., 192ff.

some two dozen bedrooms: MacGillivray (2001), 137, gives the number of bedrooms at twenty-two; Horwitz (1981), 130, puts it at twenty-eight.

a sunken Roman bath: Horwitz (1981), 130.

a mosaic floor set in a labyrinth pattern: Ibid., 129.

two huge replicas, carved in mahogany: Ibid.

"Evans' friends variously described Youlbury": Ibid.

77 *Completed in 1906*: Harden (1983), 19.

The villa's cellar was stocked: Horwitz (1981), 175.

the financier J. P. Morgan and the novelist Edith Wharton: Ibid., 204.

"On the hottest days of a Cretan summer": Ibid., 5.

"Minos," as Evans suspected: Evans (1909), iv–v, note 1.

Queen's Megaron: MacGillivray (2001), 216ff.

Domestic Quarter: Horwitz (1981), 140.

a visiting Isadora Duncan danced: James S. Candy, *A Tapestry of Life: An Autobiography* (Braunton, UK: Merlin Books,

1984), 26. Candy was the tenant farmer's son whom Evans took in as a ward. For the date, see Cathy Gere, *Knossos and the Prophets of Modernism* (Chicago: University of Chicago Press, 2009), 94.

78 *Evans spent decades clearing rubble*: See, e.g., MacGillivray (2001), 232ff.

newer materials like reinforced concrete: Gere (2009), 1.

long before early Hellenic peoples: Chadwick (1976), 4ff.

this one came from Evans himself: Evans (1909), 1ff.

superior to them in every conceivable way: Robinson (2002), 10, 33.

In his earliest writings on the Cretan scripts: See, e.g., Evans (1894), passim.

79 *Evans became convinced that the civilization he had unearthed*: Horwitz (1981), 2.

"As excavation went on": Myres (1941), 949.

He called his island culture Minoan: Evans (1921), v.

a 2002 BBC television documentary: This is *A Very English Genius*, first broadcast on the BBC in 2002.

80 *the language of Linear B was the indigenous Minoan tongue*: Evans (1894), 353ff.

The few scholars who dared to question him: Chadwick (1967), 25.

Candidates ranged from the preposterous: See, e.g., Ibid., 26ff.

"Evans does not . . . seem to have had": Ibid., 17.

81 *"The throne, , is high-backed"*: Evans (1935), 687.

"as an ideogram, and with a determinative meaning": Ibid.

"it surely indicates a royal owner": Ibid.

"its inclusion at any rate suggests": Ibid., 701.

"It is itself apparently the derivative": Ibid., 708.

82 *a crucial clue hidden in this fragmentary tablet*: Ibid., 799, note 3.

"the discoverer of the script did not achieve": Myres (1941), 953.

CHAPTER FOUR: AMERICAN CHAMPOLLION

85 *On the evening of June 15*: AEK gives the date of the lecture in a c. 1947 curriculum vitae, 4, AEK Papers, PASP.

She was by nature self-contained: AEK's former student Eva Brann describes her thus in an unpublished biographical essay, "In Memoriam: Alice E. Kober" (2005), 3–5, AEK Papers, PASP.

speaking in public made her unbearably nervous: AEK letter to JFD, Sept. 8, 1947, AEK Papers, PASP.

she typically put each of her published papers through a good ten drafts: Kober's doing so is widely attested in her correspondence and other papers. See, e.g., AEK letter to JFD, Oct. 18, 1946, in which she says, "I usually write an article over about ten times" and AEK letter to JFD, Oct. 8, 1947 (both, AEK Papers, PASP), in which she speaks of writing the fourth draft of an article (several more drafts will follow), adding, "I always discard the first ten pages of anything I write."

Before her now was her typescript: Alice E. Kober, untitled Phi Beta Kappa lecture, Hunter College (June 15, 1946); unpublished manuscript, AEK Papers, PASP.

Physically, she was unprepossessing: Brann (2005), 15.

86 *"On every kind of writing material known to man"*: Kober, untitled Phi Beta Kappa lecture, Hunter College (June 15, 1946), 1, AEK Papers, PASP.

a cumbersome load of classes, as many as five at a time: See, e.g., AEK letter to JLM, Nov. 28, 1948, AEK Papers, PASP, in which she writes, "Next week . . . I'm getting examinations from all five of my classes—130 long papers."

87 *"the person on whom an astute bettor"*: Thomas G. Palaima, "Alice Elizabeth Kober," unpublished manuscript, University of Texas, Austin (n.d.).

87 *one good article a year*: AEK letter to JFD, July 1, 1948, AEK Papers, PASP.

88 *noted together in 1927*: Cowley (1927).

a discovery previously attributed to Ventris: See, e.g., Robinson (2002), 87–88.

"There is no certain clue to the language": Alice E. Kober, "Form Without Meaning," lecture to Yale Linguistics Club (May 3, 1948), 4, unpublished manuscript, AEK Papers, PASP. Italics added.

"To get further, it is necessary to develop a science of graphics": Ibid., 12.

she declared she would make the Minoan scripts her lifework: E. Adelaide Hahn, "Alice E. Kober," obituary note, *Language* 26:3 (1950), 442.

89 *But in the coming years, on her own time*: All coursework per AEK curriculum vitae (c. 1947), PASP.

"One can remain sure that no Champollion": Quoted in Robinson (1995), 115.

ever-present cigarette at hand: Brann (2005), 4, speaks of AEK's being a chain smoker.

"working hundreds of hours with a slide-rule": Kober, "Form Without Meaning," 13.

who savored detective stories: AEK refers several times in her correspondence to reading detective stories as a pastime; see, e.g., AEK letter to JFD, Dec. 5, 1947, AEK Papers, PASP.

"method and order": This was a favorite expression of Agatha Christie's great detective Hercule Poirot. It was clear that AEK read Christie; in a published article not related to the Minoan scripts, "Tiberius, Master Detective," *Classical Outlook* 22:4 (Jan. 1945), 37, she writes of Emperor Tiberius having used his "little grey cells"—another favorite term of Poirot's—to resolve a murder investigation.

90 *Alice Elizabeth Kober was born*: Curiously, AEK's birth certificate, New York City Department of Health No. 1162, AEK Papers, PASP, lists her given name as Adele; this name appears nowhere else in her records.

Her parents had come to the United States: Passenger records, SS *Statendam*, May 29, 1906, Statue of Liberty–Ellis Island Foundation, AEK Papers, PASP.

The couple settled in Yorkville: AEK's birth certificate lists the family's address as 247 East 77th Street in Manhattan.

Census records list Franz's occupation: United States census, Bronx, NY, 1930, AEK Papers, PASP; Franz Kober death certificate, New York City Department of Health No. AA37162, AEK Papers, PASP.

In the summer of 1924, she placed third: "Awards of University Scholarships," *New York Times,* Aug. 22, 1924, 16.

she took part in the Classical Club and the German Club: Hunter College Yearbook (1928) 46, AEK Papers, PASP.

"As an undergraduate she impressed me": Ernst Reiss, letter of reference accompanying AEK's application for a grant from the John Simon Guggenheim Memorial Foundation, Nov. 1945, AEK Papers, PASP.

91 *In 1928, she was elected to Phi Beta Kappa*: "843 to Get Degrees at Hunter College," *New York Times,* June 14, 1928, 24.

she graduated magna cum laude: Ibid.

a major in Latin and a minor in Greek: AEK Hunter College transcript.

C's and D's in gym: Ibid.

followed by a Ph.D. in classics: Alice Elizabeth Kober, "The Use of Color Terms in the Greek Poets, Including All the Poets from Homer to 146 B.C. Except the Epigrammatists," Ph.D. dissertation, Columbia University, 1932.

91 *she was working the entire time*: Job history per AEK's Guggenheim grant application, Nov. 1945, 2, AEK Papers, PASP.

at an annual salary of $2,148: Salary history per ibid., 3.

"dry, refraining rigor": Brann (2005), 7.

"She was, to coin a phrase": Ibid., 15.

92 *"Everybody seems to handle Hrozný"*: AEK letter to JFD, Sept. 22, 1947, AEK Papers, PASP.

"When you wrote me in May, 1946": Cover letter accompanying AEK paper "The Language (or Languages) of the Minoan Scripts," Classical Association of the Atlantic States (May 1946), unpublished manuscript, AEK Papers, PASP. Italics added.

Her life was her work: Brann (2005) and Palaima and Trombley (2003) make a similar point.

After her father's death from stomach cancer: Franz Kober died on Dec. 17, 1935, at sixty-two. Death certificate, AEK Papers, PASP.

in the house Alice owned: In several places in her correspondence, AEK speaks of owning a house in Flatbush. See, e.g., AEK letter to JFD, Dec. 5, 1947, AEK Papers, PASP.

93 *Kober did not write or lecture about the script publicly*: The first example of scholarly work by AEK on the subject in the AEK Papers, PASP archives is her 1941 paper, "Some Comments on a Minoan Inscription (Linear Class B)," presented at a meeting of the Archaeological Institute of America (Dec. 31, 1941), AEK Papers, PASP. An abstract of the paper was published the next year in the *American Journal of Archaeology* 46:1 (1942), 124.

"I have been working on the problems presented": AEK letter to Mary H. Swindler, Jan. 29, 1941, AEK Papers, PASP.

"of the kind so successfully used": Ibid., 12.

94 *"As you are aware," he says*: Conan Doyle (1903), 522.
when Kober first turned her attention to the Cretan scripts: In a 1941 letter to Mary Swindler, editor of the *American Journal of Archaeology,* AEK writes of having worked on the Minoan scripts "for about ten years now." Jan. 29, 1941, AEK Papers, PASP.
"thesaurus absconditus": AEK to ELB, June 7, 1948, ELB Papers, PASP.

95 *linguistic survivals like* wine: All examples of linguistic survivals in British English are from Thomas Pyles, *The Origins and Development of the English Language,* 2nd ed. (New York: Harcourt Brace Jovanovich, 1971), 313ff.
"would be of great importance to scholars": Alice E. Kober, "The Gender of Nouns Ending in -inthos," *American Journal of Philology* 63:3 (1942), 320–27.

96 *pre-Hellenic words ran "into the thousands"*: AEK to JFD, Feb. 1, 1947, AEK Papers, PASP.
Among the pre-Hellenic words Kober identified: Ibid., 321.
She had tried several times to tear herself away: AEK to JFD, May 22, 1942, AEK Papers, PASP.
"I've resigned myself": Ibid.

97 *who quickly appropriated Evans's Villa Ariadne*: Horwitz (1981), 247.
"No archaeologist, however able": Alice E. Kober, "Some Comments on a Minoan Inscription (Linear Class B)," paper presented at meeting of the Archaeological Institute of America (Dec. 31, 1941); abstract published in *American Journal of Archaeology* 46:1, 124.

98 *This was also the case for the words "boy" and "girl"*: Cowley (1927).
almost assuredly meant "total": See, e.g., Alice E. Kober, "The Cryptograms of Crete," *Classical Outlook* 22:8 (May 1945), 78.

98 *Blissymbolics was invented after World War II*: See, e.g., Charles Kasiel Bliss, *Semantography: A Non-Alphabetical Symbol Writing, Readable in All Languages; a Practical Tool for General International Communication, Especially in Science, Industry, Commerce, Traffic, etc., and for Semantical Education, Based on the Principles of Ideographic Writing and Chemical Symbolism*, 3 vols. (Sydney: Institute for Semantography, 1949).

99 *the full solution*: The answer to the Blissymbolics problem appears below:

	Part of Speech	Composition	Meaning
	verb	nose + mouth	to breathe
	noun	water + mouth	saliva
	adjective	circle (sun) + pointer	western
	adjective	activity	active
	noun	body (torso) + 2 pointers	waist
	verb	mouth + (air + outwards)	to blow
	adjective	sick	sick
	noun	mouth + 2 pointers	lips
	verb	eye + (water + downwards)	to cry
	noun	activity	activity
	adjective	heart + upwards	merry

103 *"It is possible to prove, quite logically"*: Kober, "Form Without Meaning," 3–4.

"I am interested": AEK to JLM, Oct. 29, 1948, AEK Papers, PASP.

"We cannot speak of language, *but only of* script*"*: Alice E. Kober, "The Cretan Scripts," lecture to the New York Clas-

sical Club (May 17, 1947); unpublished manuscript, AEK Papers, PASP; emphasis in original.

104 *"in patterns of selection and arrangement"*: Barber (1974), 145. In making this statement, Barber credits Paul L. Garvin, *On Linguistic Method: Selected Papers* (The Hague: Mouton, 1964), 22ff., 78–79.

"each sign bears a relation": Barber (1974), 117.

she would fill forty of them: All forty notebooks are contained in the AEK Papers, PASP archive.

106 *"a distinctive fingerprint"*: Barber (1974), 18.

107 *"may be represented by from two"*: Ibid., 108.

"You can figure out for yourself": AEK to JFD, Oct. 27, 1947, AEK Papers, PASP.

108 *by stacking two or more cards together*: Kober explains the principle behind the cards in a letter to JS, Feb. 25, 1947, AEK Papers, PASP.

"Making all these files takes time": AEK to HAM, Jan. 27, 1946, AEK Papers, PASP.

"reveal a gentler side": Palaima and Trombley (2003).

"I . . . teach in my spare time": AEK letter to JFD, May 4, 1942, AEK Papers, PASP.

"Brooklyn College never did anything for me": AEK to JFD, July 1, 1948, AEK Papers, PASP.

109 *She shared an office with four other people*: AEK to ELB, Feb. 16, 1949, AEK Papers, PASP.

asked to give private instruction in Horace: AEK curriculum vitae (c. 1947), 5, AEK Papers, PASP.

110 *from 1944 onward, she brailled textbooks*: Ibid.

It took as many as fifteen hours: "Adventure by Research Proves Ample Reward," *Brooklyn Eagle*, April 28, 1946.

In a letter to her department chairman: AEK letter to "Professor Pearl" [Joseph Pearl, chairman of the Department of Classical Languages at Brooklyn College], May 21, 1946, AEK Papers, PASP.

110 *"The less said about teaching assignments, the better"*: AEK letter to ELB, Feb. 16, 1949, ELB Papers, PASP.

she established the Hunter chapter of Eta Sigma Phi: AEK letter to "Professor Pearl," March 20, 1944, AEK Papers, PASP.

111 *she took her students excavating*: AEK letter to JFD, Feb. 18, 1948, AEK Papers, PASP.

"I think I'm a good teacher": AEK letter to JFD, Oct. 27, 1947, AEK Papers, PASP.

"Dear Miss Kober": Fritzi Popper Green letter to AEK, June 27, 1944, AEK Papers, PASP.

112 *"You know a great work"*: Brann (2005), 15.

"We may find out if Helen of Troy": Kober, untitled Phi Beta Kappa lecture (June 15, 1946), 16; italics added.

CHAPTER FIVE: A DELIGHTFUL PROBLEM

113 *the newspapers announced the 132 recipients*: See, e.g., "Guggenheim Fund Makes 132 Awards," *New York Times,* April 15, 1946, 16.

"one of the 'grand old men' ": AEK curriculum vitae (c. 1947), 6.

"Your learning is great": Sir D'Arcy Wentworth Thompson letter to AEK, Aug. 25, 1946, AEK Papers, PASP.

"The very notion of your problem": Leonard Bloomfield letter to AEK, May 25, 1944, AEK Papers, PASP.

114 *"I find she is in excellent physical condition"*: Abraham L. Suchow, M.D., letter to HAM, April 8, 1946, AEK Papers, PASP.

She started work on a major review: This would become A. E. Kober (1946).

"It seems the sheerest balderdash": AEK to JFD, July 20, 1946, AEK Papers, PASP.

114 *"I hope he will not be too annoyed"*: AEK to JS, Oct. 13, 1947, AEK Papers, PASP.

"Don't cool off too much": JFD letter to AEK, Nov. 9, 1946, AEK Papers, PASP.

He and Kober began corresponding in the early 1940s: The first known letter between them is from AEK to JFD, Nov. 15, 1941, AEK Papers, PASP.

115 *She had long since mastered*: AEK undergraduate transcript, Hunter College.

From 1942 to 1945, while teaching: AEK curriculum vitae (c. 1947), AEK Papers, PASP.

"against the happy day": AEK letter to JLM, Dec. 14, 1946, AEK Papers, PASP.

"homesick for Athens": AEK to JFD, Dec. 14, 1947, AEK Papers, PASP.

She did manage a vacation: AEK letter to JFD, July 20, 1946, AEK Papers, PASP.

one of only two passing references to social life: Ibid. In the other, in a letter to JLM, July 29, 1947, AEK Papers, PASP, AEK describes having gone to the beach with friends; both she and her mother returned with "terrific sunburns."

a letter she wrote to Daniel mid-voyage: AEK letter to JFD, July 20, 1946, AEK Papers, PASP.

116 *she did them simultaneously*: AEK letter to JFD, Dec. 5, 1947, AEK Papers, PASP.

"clearing up the background": AEK letter to HAM, Sept. 8, 1947, AEK Papers, PASP.

"and some remnants of others": AEK letter to Brooklyn College president Harry D. Gideonse, Sept. 23, 1947, AEK Papers, PASP.

"It's a thankless job": AEK letter to JFD, Oct. 27, 1946, AEK Papers, PASP.

116 *"about a month of extremely intensive work"*: Ibid.

"I had never before been able to work": AEK letter to HAM, Dec. 12, 1946, AEK Papers, PASP.

117 *The letter was to Sir John Linton Myres*: Myres was knighted in 1943.

Myres had been Evans's young assistant: Horwitz (1981), 80.

"a disorganized legacy": Robinson (1995), 114.

Myres, who by the mid-1940s was elderly and ill himself: See, e.g., AEK letter to HAM, AEK Papers, PASP, in which she describes Myres as "crippled with arthritis" and "practically house-bound."

"Dear Professor Myers": AEK letter to JLM, Nov. 20, 1946, AEK Papers, PASP.

119 *"If all the Minoan scripts aren't syllabic"*: AEK to JFD, Oct. 27, 1946, AEK Papers, PASP.

120 *"form without meaning"*: Kober (1948a).

a series of tablets, reproduced by Evans in The Palace of Minos: His discussion of the "chariot" tablets can be found in Evans (1935), 786–98.

122 *In Latin, verbs can be inflected*: The verb paradigm and the noun paradigm that follows are adapted from Frederic M. Wheelock, *Latin: An Introductory Course Based on Ancient Authors* (New York: Barnes & Noble Books, 1956), 3, 8.

123 *When Evans first studied the Knossos tablets*: The first three examples are from Evans (1935), 714, the fourth from ibid., 736.

124 *He decided that these sequences were inflections*: See, e.g., ibid., 714.

125 *"We have here, surely"*: Ibid., 715.

"If a language has inflection, certain signs": Kober (1945), 143–44.

126 *"It was one thing to suggest"*: Pope (1975), 159.

127 *"When a syllabary is used [inflections] are bound to be obscured"*: Ibid., 143, note 1.

128 *On December 13, 1946, a letter with an Oxford return address*: The letter, from JLM to AEK, is dated Nov. 27, 1946, AEK Papers, PASP; for its arriving on Dec. 13, see AEK to HAM, Dec. 13, 1946, AEK Papers, PASP; for it taking an hour before Kober dared open it, see AEK to JLM, Dec. 17, 1946, AEK Papers, PASP.

"Please let me know your plans": JLM to AEK, ibid.

"My fondest hopes [*have*] *materialized"*: AEK to JLM, Dec. 17, 1946.

CHAPTER SIX: SPLITTING THE BABY

129 *"In a little over a month"*: AEK to JLM, Feb. 6, 1947, AEK Papers, PASP.

she was somewhat afraid to fly: This is amply documented throughout Kober's correspondence, e.g., AEK to JLM, Jan. 30, 1947, AEK Papers, PASP; AEK to JLM, May 1, 1947, AEK Papers, PASP.

130 *of such poor quality that it would barely take ink*: See, e.g., AEK to JLM, July 7, 1947, AEK Papers, PASP. Palaima and Trombley (2003) also raise the issue of substandard paper.

"It was quite bearable": AEK to HAM, June 23, 1947, AEK Papers, PASP.

84, Charing Cross Road: Helene Hanff, *84, Charing Cross Road* (New York: Grossman, 1970).

131 *when they began corresponding, in early 1947*: The first known letter between them is JS to AEK, Jan. 6, 1947, AEK Papers, PASP.

"Dear Miss Kober," Sundwall wrote: Ibid.

"Of all the people in the world": AEK to JS, Jan. 28, 1947, AEK Papers, PASP.

132 *"coffee in the bean"*: AEK to JS, Feb. 25, 1947, AEK Papers, PASP.

she mailed him an orange: AEK to JS, Oct. 13, 1947, AEK Papers, PASP.

132 *"Höffentlich sind Sie nicht Teetotaler"*: AEK to JS, March 12, 1948, AEK Papers, PASP.

"Mother and I decided": AEK to JS, Oct. 13, 1974, AEK Papers, PASP.

Sundwall's publication of the contraband inscriptions: Johannes Sundwall, "Minoische Rechnungsurkunden," *Societas Scientiarum Fennica, Commentationes Humanarum Litterarum* (4:4, 1932) and *Altkretische Urkundenstudien* (Åbo: Åbo Akademi, 1936).

Now, a later book of Sundwall's: Knossisches in Pylos: Johannes Sundwall, *Knossisches in Pylos* (Åbo: Åbo Akademi, 1940).

after searching for more than a year: AEK to JS, Jan. 26, 1947, AEK Papers, PASP.

"spent two very happy days": AEK to JS, Feb. 6, 1947, AEK Papers, PASP.

133 *"I've timed myself"*: AEK to JLM, Feb. 8, 1947, AEK Papers, PASP.

when her fingers were stiff with cold: Ibid.

"[Myres] mentions having a 'severe chill' ": AEK to JFD, Feb. 25, 1947, AEK Papers, PASP.

"I'll be content to copy what I can": AEK to JLM, March 6, 1947, AEK Papers, PASP.

On the seventh: AEK to JLM, Feb. 6, 1947, AEK Papers, PASP.

She planned to learn Ancient Egyptian: AEK to HAM, March 8, 1947, AEK Papers, PASP.

134 *"Everything that's been written on Minoan"*: AEK to JLM, Feb. 8, 1947, AEK Papers, PASP.

the issue that contained it: AEK to HAM, March 3, 1947, AEK Papers, PASP.

Titled "Inflection in Linear Class B": Alice E. Kober, "Inflection in Linear Class B: 1—Declension," *American Journal of Archaeology* 50:2 (1946), 268–76.

135 *"Let us suppose, for instance"*: Ibid., 150; boldface added.

136 *most Minoan words were three or four characters*: Alice E. Kober, "The Minoan Scripts: Fact and Theory," *American Journal of Archaeology* 52:1 (1948), 97.

Since many of the tablets were inventories: Kober (1946a), 269.

137 *Kober built a paradigm*: Adapted from ibid., 272.

"Kober's triplets": Robinson (2002), 69.

a hypothetical example from Latin: Adapted from Kober (1946a), 275.

140 *"If this interpretation is correct"*: Ibid., 276.

141 *"I've been devoting all the time available"*: AEK to HAM, April 2, 1947, AEK Papers, PASP.

It had taken her five years: AEK to HAM, Jan. 27, 1947, AEK Papers, PASP.

142 *"I had a most delightful time"*: AEK to JFD, April 30, 1947, AEK Papers, PASP.

"I can hardly realize I have seen him": AEK to JFD, May 6, 1948, AEK Papers, PASP.

The book was due out in early 1948: AEK to JFD, April 30, 1947, AEK Papers, PASP.

Kober left England on April 17: AEK to JLM, Jan. 20, 1947, AEK Papers, PASP.

On April 25, when the Queen Elizabeth *docked*: AEK to JFD, April 30, 1947, AEK Papers, PASP.

"I have so much to do": Ibid.

143 *Dear Mr. Moe*: AEK to HAM, Jan. 30, 1947, AEK Papers, PASP.

145 *The letter from the foundation gave no reason*: HAM to AEK, March 28, 1947, AEK Papers, PASP.

"I am in a way relieved": AEK to HAM, April 2, 1947, AEK Papers, PASP.

146 *Kober wrote a similar letter*: AEK to CWB, Nov. 20, 1946, AEK Papers, PASP.

146 *As the story went*: A précis of the story appears, e.g., in Kober (1946c).

Blegen arrived in Pylos in April 1939: William A. McDonald and Carol G. Thomas, *Progress into the Past: The Rediscovery of Mycenaean Civilization,* 2nd ed. (Bloomington: Indiana University Press, 1990), 229ff.

147 *a vault in the Bank of Greece*: Carl W. Blegen, introduction to Bennett (1951), viii.

148 *Kober's letter to Blegen*: AEK to CWB, Nov. 20, 1946, AEK Papers, PASP.

"The difficulties in the way of granting it": CWB to AEK, Dec. 9. 1946, AEK Papers, PASP.

"I am a pessimist": AEK to JLM, June 26, 1947, AEK Papers, PASP.

149 *bringing the number at her disposal*: AEK to HAM, Sept. 8, 1947, AEK Papers, PASP.

manuscripts on Minoan sent in "by crack-pots": JFD to AEK, Sept. 3, 1946, AEK Papers, PASP.

"I rather consider myself an expert": AEK to JFD, Sept. 4, 1946, AEK Papers, PASP.

CHAPTER SEVEN: THE MATRIX

151 *"in academic harness"*: AEK to JFD, Sept. 22, 1947, AEK Papers, PASP.

"It must be quite wonderful": AEK to JS, June 3, 1947, AEK Papers, PASP.

Daniel seized on the idea: JFD to AEK, Sept. 11, 1947, AEK Papers, PASP.

152 *"Compared to you a hurricane"*: AEK to JFD, Sept. 22, 1947, AEK Papers, PASP.

to be known as the Center for Minoan Linguistic Research: JFD to AEK, June 19, 1948, AEK Papers, PASP.

152 *"You are the person in this country"*: JFD to AEK, Sept. 11, 1947, AEK Papers, PAS

"I have a job which, while far from ideal": AEK to JFD, Sept. 18, 1947, AEK Papers, PASP.

"Dangling . . . the Institute in front of my nose": AEK to JFD, Sept. 22, 1947, AEK Papers, PASP.

"a big salary for a woman teacher," more than $6,000: Ibid.

"I know enough about academic salaries for women": Ibid.

153 *By chance Roland Kent*: JFD to AEK, Sept. 19, 1947, AEK Papers, PASP.

Greek phonology: JFD to AEK, Dec. 6, 1947, AEK Papers, PASP.

"Don't count on it too much": JFD to AEK, Sept. 19, 1947, AEK Papers, PASP.

"Your latest letter . . . has me sitting here": AEK to JFD, Dec. 8, 1947, AEK Papers, PASP.

154 *from 1938, "as might be expected"*: AEK to Mary Swindler, Oct. 10, 1945, AEK Papers, PASP.

she had to enlist the aid of the Czech Consulate: AEK to JFD, July 20, 1946, AEK Papers, PASP.

155 *"Everybody is interested in the Minoan script"*: JFD to AEK, Sept. 7, 1947, AEK Papers, PASP.

"One of the remarkable things": AEK to JFD, Sept. 8, 1947, AEK Papers, PASP.

"About the article": AEK to JFD, Sept. 22, 1947, AEK Papers, PASP.

hard at work on the fourth draft: AEK to JFD, Oct. 8, 1947, AEK Papers, PASP.

156 *"the sixth draft (durn it!)"*: AEK to JFD, Oct. 18, 1947, AEK Papers, PASP.

a "slight discovery": AEK to JFD, Dec. 14, 1947, AEK Papers, PASP.

156 *historians have attributed it to Ventris*: Michael Ventris, Work Note 1 (Jan. 28, 1951). In Michael Ventris, *Work Notes on Minoan Language Research and Other Unedited Papers,* edited by Anna Sacconi (Rome: Edizioni dell'Ateneo, 1988), 143.

157 Senatus Populus**que** Romanus: This example also appears in Robinson (2002), 88.

She made the discovery on December 14: In her letter to Daniel, Dec. 14, 1947, Kober writes of having made her discovery "to-day."

"Too bad I didn't discover it": Ibid.

158 *which begin that March*: The first letter between them in the AEK Papers, PASP archive is MV to AEK, March 26, 1948, MV Papers, PASP.

"I'm extremely glad to have them": MV to AEK, May 23, 1948, MV Papers, PASP.

William T. M. Forbes: His first known letter is WTMF to AEK, Oct. 19, 1945, AEK Papers, PASP.

the Minoan language was a form of Polynesian: See, e.g., WTMF to AEK, Oct. 31, 1945, AEK Papers, PASP.

"But now back to the Lepidoptera for a time": WTMF to AEK, May 1, 1947, AEK Papers, PASP.

159 *"No!" "Right!!"*: Kober's margin notes in, respectively, WTMF to AEK, Oct. 19, 1945, and April 2, 1946; AEK Papers, PASP.

"I should be very interested to hear": MV to AEK, March 26, 1948, MV Papers, PASP.

a belief he had first made public: M.G.F. Ventris, "Introducing the Minoan Language," *American Journal of Archaeology* 44:4 (1940), 494–520.

160 *"At the moment I'm engaged"*: MV to AEK, n.d., but demonstrably before May 23, 1948, MV Papers, PASP.

"Rivalry has no place in true scholarship": AEK to ELB, Feb. 12, 1949, ELB Papers, PASP.

160 *"Please send me FAST"*: JFD to AEK, Dec. 3, 1947, AEK Papers, PASP.

161 *"I am indeed astonished"*: Franklin Edgerton to AEK, Dec. 5, 1947, AEK Papers, PASP.

"Everybody tells me this job": AEK to JFD, Dec. 6, 1947, AEK Papers, PASP.

162 *Sir John Myres wanted her to help lobby*: See, e.g., AEK to JLM, Oct. 21, 1947, AEK Papers, PASP.

"First Class. Ouch!": AEK to JFD, April 24, 1948, AEK Papers, PASP.

"Six months is ample time": AEK to JFD, Sept. 18, 1947, AEK Papers, PASP.

Myres would outlive her: Myres died on March 6, 1954.

163 *"The basic distinction between fact and theory"*: Kober (1948b), 82.

164 *Its use in archaeological decipherment dates*: Pope (1975), 163.

"was the idea of constructing such a grid": Ibid.; italics added.

166 *"People often say"*: Ibid., 102.

167 *"Let us face the facts"*: Ibid., 102–3.

a "slight setback": JFD to AEK, Dec. 19, 1947, AEK Papers, PASP, for quotation; AEK to JFD, Dec. 21, 1947, for its having been a special-delivery letter.

"You are no. 2": JFD to AEK, Dec. 19, 1947, AEK Papers, PASP.

168 *"Don't you think a lot of the opposition"*: AEK to JFD, Dec. 21, 1947, AEK Papers, PASP.

"The fact that you are a woman": JFD to AEK, Dec. 24, 1947, AEK Papers, PASP.

"weak, lazy, and impressionable": Ibid.

169 *"I am not being unduly bitter"*: Ibid.

the genus professoricus [sic]: The correct form is "genus professoricum."

169 *"I am limiting my activities"*: JFD to AEK, Feb. 14, 1948, AEK Papers, PASP.

"I have bad news": JFD to AEK, May 4, 1948, AEK Papers, PASP.

"Well, it was fun while it lasted": AEK to JFD, May 6, 1948, AEK Papers, PASP.

Penn wanted to go ahead: JFD to AEK, June 2, 1948, AEK Papers, PASP.

she was named a research associate: Mrs. William S. Godfrey, Penn University Museum secretary, to AEK, July 8, 1948, AEK Papers, PASP.

170 *a "mutual aid society"*: AEK to JFD, July 1, 1948, AEK Papers, PASP.

"If it works as we hope": AEK to HAM, July 17, 1948, AEK Papers, PASP.

At the top of the guest list: AEK to JFD, July 1, 1948, AEK Papers, PASP.

The two men had begun corresponding: Robinson (2002), 44.

171 *Even his everyday handwriting*: Postcard from MV to JLM, Jan. 19, 1954, MV Papers, PASP.

"Mr. Ventris would have no trouble": AEK to JLM, July 8, 1948. Quoted in Thomas G. Palaima, Elizabeth I. Pope, and F. Kent Reilly III, *Unlocking the Secrets of Ancient Writing: The Parallel Lives of Michael Ventris and Linda Schele and the Decipherment of Mycenaean and Mayan Writing,* exhibition catalogue, Nettie Lee Benson Latin American Collection, University of Texas (Austin: Program in Aegean Scripts and Prehistory, 2000).

172 *a long voyage through Greece, Cyprus, and Turkey*: JFD to AEK, Aug. 13, 1948, AEK Papers, PASP.

to scout sites for future excavations: "J. F. Daniel 3d Dies; Archaeologist, 38," *New York Times,* Dec. 19, 1948, 76.

172 *he was due to set sail for Athens*: Ibid.

"Until it is," she wrote: AEK to HAM, July 17, 1948, AEK Papers, PASP.

Kober found Myres even frailer: AEK to JFD, Aug. 9, 1948, AEK Papers, PASP.

"Lady Myres keeps him in bed": AEK to JFD, Aug. 24, 1948, AEK Papers, PASP.

"is in perfect chaos": Ibid.

"I only hope he accepts my corrections": Ibid.

Kober understood that Myres was living : AEK to JS, Feb. 11, 1948, AEK Papers, PASP.

173 *"I've had enough trouble"*: AEK to JFD, Aug. 4, 1948, AEK Papers, PASP.

"Two deans . . . still wish": JFD to AEK, Aug. 31, 1948, AEK Papers, PASP.

174 *"He wants to use my classification"*: AEK to JFD, Aug. 4, 1948, AEK Papers, PASP.

"I am really very much in awe": AEK to JLM, Dec. 17, 1946, AEK Papers, PASP.

"still thinks . . . that there are no cases": AEK to JS, Oct. 3, 1948, AEK Papers, PASP.

"Somewhat against my better judgement": AEK to ELB, Oct. 3, 1948, ELB Papers, PASP.

175 *On September 18*: AEK to JLM, Sept. 26, 1948, AEK Papers, PASP.

"I've been home almost a month now": AEK to JLM, Oct. 14, 1948, AEK Papers, PASP.

All this left her barely an hour a day: AEK to ELB, Oct. 13, 1948, ELB Papers, PASP.

"Wishing you the best of luck": JDF to AEK, Sept. 7, 1948, AEK Papers, PASP.

On December 18, 1948: *New York Times*, Dec. 19, 1948.

CHAPTER EIGHT: "HURRY UP AND DECIPHER THE THING!"

177 *"The news came as a terrible shock"*: AEK to JLM, Dec. 26, 1948, AEK Papers, PASP.

"I did not write more about Daniel": AEK to JS, March 19, 1949, AEK Papers, PASP.

178 *"I'm neck-deep in Sir John's manuscript"*: AEK to ELB, Nov. 22, 1948, ELB Papers, PASP; italics added.

A sample of John Myres's handwriting from 1948: JLM to MV, March 7, 1948, MV Papers, PASP.

179 *"He showed slides"*: AEK to JFD, May 15, 1942, AEK Papers, PASP.

The symbols [symbol], [symbol], and [symbol], for instance: ELB to AEK, April 19, 1948, ELB Papers, PASP.

appeared at Knossos but not at Pylos: AEK postcard to ELB, May 15, 1949, ELB Papers, PASP.

"Dr. Bennett . . . is a very agreeable young man": AEK to JS, June 22, 1948, AEK Papers, PASP.

180 *Bennett, who had done his doctoral dissertation*: Emmett L. Bennett Jr., "The Minoan Linear Script from Pylos," Ph.D. dissertation, University of Cincinnati, 1947.

"Bennett suggested I get his dissertation": AEK to JFD, June 3, 1948, AEK Papers, PASP.

"There can be no doubt now": AEK to ELB, June 3, 1948, ELB Papers, PASP.

181 "If *there is a possibility"*: AEK to ELB, June 7, 1948, ELB Papers, PASP.

"Happy, happy day!": AEK to ELB, Nov. 22, 1948, ELB Papers, PASP.

"If Bennett is willing": AEK to JLM, Oct. 29, 1948, AEK Papers, PASP.

"a large, well-tabled room": ELB to AEK, May 10, 1949, ELB Papers, PASP.

182 *"the last time I did it"*: ELB to AEK, June 14, 1949, ELB Papers, PASP.

"There are a few signs that must be added": AEK to ELB, June 22, 1948, ELB Papers, PASP.

183 *In 1949, Kober came out with a noteworthy article*: Alice E. Kober, " 'Total' in Minoan (Linear Class B)," *Archiv Orientální* 17 (1949), 386–98.

186 *"I shall probably give the problem a rest"*: MV to AEK, Feb. 22, 1949, MV Papers, PASP.

"I just finished my last set of examinations": AEK to ELB, June 22, 1948, ELB Papers, PASP.

"About my errors. I must apologize": Ibid.

"I am ashamed at the number of errors": AEK to JS, Feb. 21, 1949, AEK Papers, PA

"This year has been a nightmare": AEK to JLM, May 2, 1949, AEK Papers, PASP.

an uncompromising regimen of dieting: E. Adelaide Hahn (1950), 443.

187 *"If you have the time"*: ELB to AEK, Aug. 12, 1949, ELB Papers, PASP.

with things like packaged soup: AEK to JLM, June 8, 1948, AEK Papers, PASP.

the nearest post office was more than a mile away: AEK to JS, Nov. 8, 1947, AEK Papers, PASP; AEK to JLM, June 15, 1949, AEK Papers, PASP.

"because he makes so many little errors": AEK to JS, Nov. 19, 1948, AEK Papers, PASP. Original in German; my translation.

188 *"I never know when proofs are coming"*: AEK to ELB, June 30, 1949, ELB Papers, PASP.

"School is just finishing": AEK to JLM, June 15, 1949, AEK Papers, PASP.

188 *"Sorry that at times"*: AEK to JLM, Oct. 24, 1948, AEK Papers, PASP.

"I want to get back to my own job": AEK to JLM, Nov. 7, 1948, AEK Papers, PASP.

189 *"What I would like to do right now"*: AEK to JLM, Nov. 10, 1948, AEK Papers, PASP.

"Now, my reason for this letter": AEK to JLM, Nov. 28, 1948, AEK Papers, PASP.

190 *In early July 1949*: AEK to JS, AEK Papers, PASP.

On July 27, she was ordered to the hospital: AEK to JLM Aug. 30, 1949, AEK Papers, PASP.

"I'm sorry," she told Bennett: Ibid.

"I managed to acquire something": AEK to JLM, Nov. 7, 1949, AEK Papers, PASP.

191 *it was whispered among the women in the family*: Patricia Graf, personal communication.

It has also been suggested: Hahn (1950).

In a short letter from late August: AEK to JS, Aug. 31, 1949, AEK Papers PASP.

"It is too bad": AEK to JLM, Aug. 30, 1949, AEK Papers, PASP.

192 *a second, six-week hospital stay*: AEK to ELB, Oct. 29, 1949, ELB Papers, PASP.

"Professor Blegen at last relented": AEK to HAM, Oct. 29, 1949, AEK Papers, PASP.

"I hope I am now": AEK to ELB, Oct. 29, 1949, ELB Papers, PASP.

By return mail: ELB to AEK, Nov. 1, 1949, ELB Papers, PASP.

Kober was officially on sick leave: AEK to JLM, Nov. 7, 1949, AEK Papers, PASP.

"My health is, unfortunately": Ibid.

192 *"I haven't done anything about going to Greece"*: Ibid.

195 *Brooklyn College promoted her*: Harry D. Gideonse to AEK, Jan. 19, 1950, AEK Papers, PASP.

Writing to Myres that month: AEK to JLM, Feb. 18, 1950, AEK Papers, PASP.

In a brief, scrawled letter to Sundwall: AEK to JS, March 4, 1950, AEK Papers, PASP.

196 *That day, in a postcard to Bennett*: AEK postcard to ELB, March 4, 1950, ELB Papers, PASP.

"still busy checking": AEK postcard to ELB, April 4, 1950, ELB Papers, PASP.

That spring, a long article by Kober: This is Kober (1950).

her last publication: A bibliography of Kober's work appears in Sterling Dow, "Minoan Writing," *American Journal of Archaeology*, 58:2 (1954), 83–84.

"In the ultimate analysis": Kober (1950), 293–95; italics added.

197 *Kober wrote an astonishing letter to Myres*: AEK to JLM, April 17, 1950, AEK Papers, PASP.

199 *On the morning of May 16, 1950*: AEK death certificate (No. 156-50 310216, New York City Department of Health); "Prof. Alice Kober of Brooklyn Staff," obituary, *New York Times*, May 17, 1950, 29.

CHAPTER NINE: THE HOLLOW BOY

203 *plus one younger schoolmate*: Leonard Cottrell, "Michael Ventris and His Achievement," *Antioch Review* 25:1 (1965), 14.

204 *"Did you say the tablets"*: Robinson (2002), 21.

205 *Edward Francis Vereker Ventris*: Marjorie Dent Candee, ed., *Current Biography Yearbook* (New York: H. W. Wilson, 1958), 566.

born in Wheathampstead: Cottrell (1965), 13.

205 *On his father's side, Ventris was descended*: Robinson (2002), 16.

His paternal grandfather: Robinson (2002b), 16.

206 *"Overshadowed by illness"*: Ibid.

numbered among her friends: Ibid., 29; Palaima, Pope, and Reilly (2000), 7.

Edward Ventris suffered from tuberculosis: Robinson (2002), 17.

He attended a boarding school in Gstaad: Ibid.; Cottrell (1965), 14.

French, German, and the local Swiss German dialect: Robinson (2002).

he bought and devoured Die Hieroglyphen: Palaima, Pope, and Reilly (2000), 7.

much has been made of his prodigious ability: See, e.g., Cottrell (1965), 14; *A Very English Genius* (2002).

207 *the "critical period" for language acquisition*: See, e.g., Margalit Fox, *Talking Hands: What Sign Language Reveals About the Mind* (New York: Simon & Schuster, 2007), 125ff.

"How do you come to be so expert": Robinson (2002), 117.

208 *"I heard it all"*: Jean Overton Fuller to Andrew Robinson, May 2002, MV Papers, PASP.

209 *Jung, Fuller wrote, "had complicated matters"*: Ibid.

"I think they rather thought me": MV to LV, World War II–era letter dated only "Wednesday night" [1942], MV Papers, PASP.

a middling student: Robinson (2002); *A Very English Genius* (2002).

night after night, after lights-out: *A Very English Genius* (2002).

From about 1932 on: Ibid.

210 *The couple formally divorced*: Robinson (2002), 26; *A Very English Genius* (2002).

210 *"I count myself extraordinarily fortunate"*: Quoted in Robinson (2002), 28–29.

211 *"Dear Sir," begins one letter*: MV to AJE, Easter 1940, MV Papers, PASP.

212 *He would discreetly neglect to tell the editors*: Cottrell (1965), 19.

forced to withdraw Michael from Stowe: Robinson (2002), 30; *A Very English Genius* (2002).

He wrote to his mother's friend Marcel Breuer: Robinson (2002), 30.

Ventris enrolled there in January 1940: Ibid., 31.

"She had already lost her brother": Ibid.

"The coroner's verdict": Ibid., 40.

213 *To the end of his life, Robinson added*: Ibid.

He tore up two drafts: Ibid., 32.

"Dear Sir: I am enclosing an article": MV to *American Journal of Archaeology*, Sept. 22, 1940, MV Papers, PASP.

214 *"close to worthless"*: Palaima (n.d.), 17.

she did so as an act of mercy: Ibid.

a classmate a few years older than he: Robinson (2002), 42.

"It looks as if, in the ordinary way": Ibid., 42–43.

215 *Ventris and Lois married*: Ibid., 43.

"the nicest present": Quoted in Robinson (2002b), 45.

Called up in the summer of 1942: Robinson (2002), 43.

"Darling Lois": MV to LV, World War II–era letter dated only "Scarborough, Sunday" [probably late 1942], MV Papers, PASP.

217 *"My knowledge is gradually getting on"*: MV to LV, World War II–era letter dated only "Wednesday night" [1942], MV Papers, PASP.

it interested him far more than actual flying did: Cottrell (1965), 18.

217 *"It's a desk job, really"*: *A Very English Genius* (2002).

"on one occasion he horrified his captain": Cottrell (1965), 18–19.

218 *because of his foreign-language prowess*: Robinson (2002), 47.

a daughter, Tessa, had been born: Ibid.

He also met with Myres: Ibid.; JLM to MV, April 23, 1948; MV to JLM, May 6, 1948, MV Papers, PASP.

Ventris and Lois graduated with honors: Robinson (2002), 57.

219 *"Dear Sir John"*: MV to JLM, dated only "Monday night" [1948], MV Papers, PASP.

220 *he was immensely pleased*: See, e.g., JLM to MV, March 7, 1948, MV Papers, PASP.

"The man who may decipher Linear B": Cottrell (1965), 29.

CHAPTER TEN: A LEAP OF FAITH

221 *In September 1949*: Robinson (2002b), 73.

"He was," she wrote: Prue Smith, *The Morning Light: A South African Childhood Revalued* (Cape Town: David Philip, 2000), 239.

"a strange architectural drawing aid": Ibid.

222 *"with instant accuracy"*: Ibid.

working feverishly on the script during lunch breaks: Robinson (2002b), 74.

"It is hard to see": Ibid.

translating the replies as needed: Ventris (1988), 32.

At his own considerable expense: Ibid.

"I have good hopes": Ibid., 108.

223 *Then, in the summer of 1950*: Robinson (2002b), 76.

made out handsomely in the stock market: Ibid., 74.

soon quit his job: Ibid., 77.

224 *Work Note 1*: Ventris (1988), 135ff.

Ventris independently replicates: Ibid., 143.

224 *his own first attempt on paper at a phonetic grid*: Ibid.

Ventris had previously built a three-dimensional "grid": Robinson (2002b), 81.

225 *"must be regarded as a failure"*: Pope (1975), 164.

"a suspicious . . . official asked him": Robinson (2002b), 83.

the first established signary for the script: Bennett (1951), 82.

227 *His signary looked much like this*: As depicted here, the signary includes slight modifications based on later findings. After Bennett (1951), 82; Robinson (2002a), 88.

229 *"represent the values which seem the most useful"*: Ventris (1988), 315.

Even in his third grid: Pope (1975), 164.

230 *"I must say I was slightly disappointed"*: MV to JLM, Jan. 28, 1948, MV Papers, PASP.

"alternative name-endings": MV to AEK, Good Friday [March 26], 1948, MV Papers, PASP.

including [illegible] *and* [illegible]: Pope (1975), 166.

"which on the face of it": MV to JLM, Aug. 28, 1951, MV Papers, PASP.

a conference on the ancient Near East: MV to JLM, Sept. 11, 1951, MV Papers, PASP.

"I was frankly rather disappointed": MV to JLM, Oct. 5, 1951, MV Papers, PASP.

231 *The words unique to Knossos included*: Adapted from Kober (1946), 274.

Perhaps, Ventris conjectured in early 1952: MV to JLM, Feb. 28, 1959, MV Papers, PASP.

232 *"only a little adjustment"*: Ibid.

"the theory that Minoan could be Greek": Ventris (1940), 494.

CHAPTER ELEVEN: "I KNOW IT, I *KNOW* IT"

233 *"all your help in the B volume"*: JLM to AEK, Oct. 26, 1948, AEK Papers, PASP.

the published book gave scant indication: Evans (1952), vi. Myres's acknowledgment reads as follows: "Thanks are due . . . to Dr. Alice Kober, of Brooklyn College, New York, who came twice to Oxford to study the unpublished texts, revised the Vocabulary, contributed to the Inventory of tablets according to their contents, read the proofs, and contributed many valuable suggestions. She was ready to go also to Crete, if the Candia Museum had been restored so as to make the original tablets accessible."

between about the seventh and the second centuries B.C.: Pope (1975), 123.

234 *A handful of Cypriot characters looked like Linear B signs*: Adapted from Cowley (1927).

235 *"had no results"*: AEK to JLM, Sept. 18, 1947, AEK Papers, PASP.

236 *Evans tried substituting Cypriot values*: Evans (1935), 799.

"those who believe that the Minoan Cretans": Ibid., note 3.

243 do-we-lo-se: Pope (1975), 170.

[illegible] *did not appear at the ends of Linear B words*: Ibid.

244 *Ventris pinpointed additional words*: The examples are from Ventris (1988), 337ff.

"In the chains of deduction": Ibid., 327.

245 *"These may well turn out"*: Ibid.

"Lois Ventris, whom we always called Betts": Smith (2000), 240. In her memoir, Smith spells Lois's nickname "Bets," but it is clear from Ventris's own correspondence (e.g., MV to ELB, May 31, 1955, PASP) that the correct spelling was "Betts."

CHAPTER TWELVE: SOLUTION, DISSOLUTION

247 *after reworking his script several times*: Robinson (2002b), 106.

The broadcast, as Andrew Robinson notes: Ibid., 104–5.

high, light, cultured, melodious: *A Very English Genius* (2002).

"For half a century, [the] Knossos tablets": Michael Ventris, "Deciphering Europe's Earliest Scripts," text of BBC Radio talk, first broadcast July 1, 1952. In Ventris (1988), 363–67.

248 *Chronologically, the Greek dialect they contained*: Michael Ventris and John Chadwick, "Evidence for Greek Dialect in the Mycenaean Archives," *Journal of Hellenic Studies* 73 (1953), 90.

249 *Even Bennett and Myres remained unpersuaded*: Robinson (2002b), 106.

"a frivolous digression": Ventris (1988), 327.

He got in touch with Myres: Robinson (2002b), 111.

"He sat as usual in his canvas chair": Chadwick (1958), 69.

250 *"I think we must accept the fact"*: Robinson (2002b), 111.

"Dear Dr. Ventris": JC to MV, July 13, 1952; in Ventris (1988), 352–53.

251 *in the Greek of Homer's time, "the" was a rarity*: Chadwick (1958), 70.

playing the dogged Watson: Robinson (2002b), 14.

"Evidence for Greek Dialect in the Mycenaean Archives": Ventris and Chadwick (1953), 84–103.

Another, published in the British journal Antiquity: John Chadwick, "Greek Records in the Minoan Script," *Antiquity* 108 (1953), 196–206; includes "A Note on Decipherment Methods," by Michael Ventris.

They also began work on a massive book: Michael Ventris and John Chadwick, *Documents in Mycenaean Greek: Three Hundred Selected Tablets from Knossos, Pylos and Mycenae with*

Commentary and Vocabulary (Cambridge: Cambridge University Press, 1956).

251 *"had attacks of cold feet"*: Chadwick (1958), 70.

"Every other day I get so doubtful": Quoted in Ibid.

"I feel it would be appropriate": Quoted in Robinson (2002b), 117.

252 *The character* [Linear B character], *for instance*: Ibid., 128.

known officially by the unromantic name P641: This drawing of the "tripod tablet," made by Ventris, was first published in his article "King Nestor's Four-Handled Cups: Greek Inventories in the Minoan Script," *Archaeology* 7:1 (Spring 1954), 18.

254 *"tripod cauldron(s) of Cretan workmanship"*: Robinson (2002b), 119.

"Is coincidence excluded?": Quoted in Chadwick (1958), 81.

the "great state of excitement": Ibid.

"Looks hard to beat!": Robinson (2002b), 121.

255 *On seeing it, the audience*: Chadwick (1958), 88.

It "went off all right": Ibid.

He spoke before the king of Sweden: Robinson (2002b), 131.

He spoke at Oxford: Ibid., 117.

He spoke at Cambridge: Ibid.

the Times *of London carried an account*: *Times*, June 25, 1953, 1.

"the Everest of Greek archaeology": Robinson (2002b), 122.

"Whichever is regarded as the greatest": Pope (1975), 9.

256 *"the label 'Minoan' had been out of date"*: Chadwick (1958), 73.

"Offers to join the academic world": Robinson (2002b), 137.

his mere three years of schoolboy Greek: Thomas G. Palaima, personal communication.

257 *"After the* Times *article"*: Robinson (202b), 126.

"By 1956, after fourteen years": Ibid., 148–49.

258 *In July 1955*: Ibid., 140.

"I shan't be able to devote": Ibid., 141.

"Information for the Architect": Ibid., 142.

259 *"There were almost no books"*: Ibid.

"One might ring up": Quoted in Ibid., 145–46.

260 *"that he himself saw no future"*: Ibid., 147.

"was aware," Robinson writes: Ibid., 148.

an "extraordinary, shocking, abject, private letter": Ibid., 149.

"I have had a couple of weeks": Quoted in Ibid., 149–50.

261 *Very late at night*: Ibid., 151.

He apparently told his family: *A Very English Genius* (2002).

He collided with a parked truck and was killed instantly: Coroner's inquisition, Hertford district of Hertfordshire, Sept. 11, 1956, MV Papers, PASP.

262 *At the coroner's inquest*: Ibid.; Robinson (2002b), 151.

"I don't think he committed suicide": *A Very English Genius* (2002).

263 *"series of fundamental articles"*: Michael Ventris, "The Decipherment of the Mycenaean Script," *Proceedings of the Second International Congress of Classical Studies* (Copenhagen: Ejnar Munksgaard, 1958), 72.

264 *"was probably too restrained"*: Robinson (2002b), 72.

"had no results": AEK to JLM, Sept. 18, 1947, AEK Papers, PASP.

"Every one of Kober's inferences": Pope (1975), 162.

265 *"alternative name-endings"*: MV to AEK, Good Friday [March 26], 1948, MV Papers, PASP.

266 *In the margin of his letter*: WTMF to AEK, May 1, 1947, AEK Papers, PASP.

267 *a phonetic grid containing more than twenty Linear B characters*: Thomas G. Palaima, "Scribes, Scribal Hands and Palaeography," in Yves Duhoux and Anna Morpurgo Davies,

eds., *A Companion to Linear B: Mycenaean Texts and Their World,* vol. 2 (Louvain-la-Neuve: Peeters, 2011), 51, note 29.

EPILOGUE: MR. X AND MR. Y

269 *"we may only find out"*: Kober, untitled lecture (June 15, 1946), 16.

"As for what the humanities": Robinson (2002b), 157.

270 *"The Criminal Courts can only tell us"*: Murray Kempton, "When Constabulary Duty's to Be Done," *New York Newsday* (May 11, 1990).

"the movement of goods": Cynthia W. Shelmerdine, "Mycenaean Society," in Yves Duhoux and Anna Morpurgo Davies, eds., *A Companion to Linear B: Mycenaean Texts and Their World,* vol. 1 (Louvain-la-Neuve: Peeters, 2008), 115.

271 *"Almost all parts of Greece"*: J. L. García Ramón, "Mycenaean Onomastics," in Duhoux and Morpurgo Davies (2011), 242.

their combined population: Shelmerdine (2008), 136.

about forty Linear B tablets were uncovered: Robinson (2002b), 117–18.

as recently as 2010: John Noble Wilford, "Greek Tablet May Shed Light on Early Bureaucratic Practices," *New York Times,* April 5, 2011, D3.

272 *Scholars have conjectured*: Palaima (2011), 116ff.

"Mycenaean state bureaucracy": Shelmerdine (2008), 127.

"His status," Shelmerdine explains: Ibid., 128.

273 *As Shelmerdine points out*: Ibid., 128–29.

rowers, as well as smallholders: Ibid., 130.

hekwetai: Ibid., 131.

"collectors": Ibid., 132.

the scribes themselves: Palaima, "Scribes, Scribal Hands and Palaeography," in Duhoux and Morpurgo Davies (2011), 121ff.

At the regional level: Shelmerdine (2008), 133–34.

274 *records the acquisition of a slave*: Yves Duhoux, "Mycenaean Anthology," in Duhoux and Morpurgo Davies (2008), 252.

rations of grain (wheat or barley), figs, and bedding: See, e.g., Shelmerdine (2008), 122, 138, 147.

"The tablets reinforce the view": Ibid., 138.

275 *"Slave status is suggested"*: Ibid., 139.

one such group is described as "captives": Ibid.

"are identified by [non-Greek] ethnic adjectives": Ibid.

"Gladly Welcome": García Ramón (2011), 220ff.

276 *while "highly expressive"*: Ibid., 226.

"Goat-Head": Ibid., 226–27.

the premonetary society of Mycenae: See, e.g., Shelmerdine (2008), 145.

A group of eight hundred tablets from Knossos: Chadwick (1976) 127.

the names of individual oxen: García Ramón (2011), 229; Chadwick (1958), 119.

At Knossos, an office in the east wing: Shelmerdine (2008), 127.

277 *"The most extraordinary figure for wheat"*: Chadwick (1976), 117–18.

"The absence of any record of the grain harvest": Ibid., 118.

The central palaces exacted payment: See, e.g., Shelmerdine (2008), 145ff.; J. T. Killen, "Mycenaean Economy," in Duhoux and Morpurgo Davies (2008), 189–90.

278 *members of certain professions, including bronzesmiths*: Shelmerdine (2008), 146.

Four metals are mentioned: Bernabé and Luján (2008), 227.

"used for a variety of purposes": Ibid.

the heads of spears and javelins: Ibid., 216.

Such spearheads, the two scholars point out: Ibid.

the nine-legged tables: Ibid., 202.

often inlaid with ivory: Ibid., 203.

279 *recorded as being made of ebony*: Ibid., 204.
Drawn by two horses: Ibid., 206.
"We are well informed": Ibid.
"is a delivery record concerning": Ibid., 207.
Woolens are sometimes described: Ibid., 218.
dyed in hues of purple and red: Ibid., 218–19.

280 *One type, pharweha*: Ibid., 219.
"Remarkable at Pylos": Shelmerdine (2008), 118.
"Hundreds of drinking and eating vessels": Ibid.
the production, transport, and delivery: Peter G. van Alfen, "The Linear B Inscribed Vases," in Duhoux and Morpurgo Davies (2008), 238.
"The production of perfumes": Bernabé and Luján (2008), 227.

281 *the ointment-maker infused wine with spices*: Ibid., 228–30.
cloth and perfumed oil were also traded overseas: Killen (2008), 184ff.
"Minoan society in Crete": Chadwick (1976), 159.
This included suits of armor: Bernabé and Luján (2008), 213ff.

282 *"Thus the watchers are guarding"*: Chadwick (1976), 175.
Another lists eight hundred rowers: Ibid.
Similar conscription records: Shelmerdine (2008), 147.
"The Linear B documents concern": Stefan Hiller, "Mycenaean Religion and Cult," in Duhoux and Morpurgo Davies (2011), 170.
"striking proof of a high degree": Hiller (2011), 205.
like Dionysus: Ibid., 183ff.

283 *beginning with the word* potnia: Ibid., 187ff.; García Ramón (2011), 235.
They include Posidaeia: Hiller (2011), 187.
twenty-two linen cloths: Ibid., 175.
manufactured items like gold vessels: Hiller (2011), 174.

283 *"3 bulls are sent"*: Ibid., 176–77.

"We know that from Homer onwards": Ibid., 177–78.

284 *from a mural in the Pylos palace*: Shelmerdine (2008), 128.

"The book-keeping testifies": Hiller (2011), 203.

"What actually happened": Chadwick (1976), 177–78.

ACKNOWLEDGMENTS

The Riddle of the Labyrinth would not exist without the energy and generosity of Thomas G. Palaima of the University of Texas. One of the world's foremost experts on Mycenaean Greece (his work earned him a MacArthur Fellowship in 1985), he has been more instrumental than anyone else in bringing Alice Kober's role in the decipherment of Linear B to light. The archives of the Program in Aegean Scripts and Prehistory (www.utexas.edu/research/pasp)—the repository Tom established at Texas and continues to superintend—have furnished a much-needed home for Kober's papers, as well as housing the papers of Tom's mentor, Emmett L. Bennett Jr., and material pertaining to Michael Ventris.

It was Tom who first put me on to Alice. Several years ago, thinking I wanted to write a book about Linear B (and thinking, as everyone did then, that the story belonged exclusively to Ventris), I cold-called him at his office. It was enough of a boon that Tom invited me, sight unseen, to come to Austin and troll through the archives. But the most valuable gift he gave me—for by divine providence, the cataloguing of Kober's papers had been completed shortly before I called—was to make me understand that the real reason to visit, and the real reason to write a book on the decipherment, was to bring the work of this fascinating, vital, yet too-long-overlooked woman into public view at last.

Others at the University of Texas have also been supremely helpful, among them Christy Costlow Moilanen, who created the meticulous finding aids to the Kober papers; the delightful Beth

Chichester, whose digital photography skill transformed images from the archive into publishable shape; Matthew Ervin; Brandi Buckler; Dygo Tosa; and Zachary Fischer. My friend Kathleen McElroy, who despite the valiant efforts of the entire newsroom to restrain her, left a high-level editing job at the *New York Times* to get a Ph.D. at U. Texas, provided aid and comfort to a weary traveler in Austin.

Although we have never met, I also owe a debt to the Scottish writer Alison Fell. In the course of researching her 2012 novel, *The Element -inth in Greek* (Sandstone Press), in which Alice Kober appears as a character, she scoured the United States for papers relating to the Kober family—birth and death certificates, ships' manifests, photographs, and the like—with which she generously seeded the archives at U. Texas. Kober's cousin Patricia Graf also graciously spent time on the phone with me, answering my questions about "Aunt Alice" and the Kober family.

I am indebted, as always, to Mark Aronoff of Stony Brook University, my longtime adviser on matters linguistic; and to Paul McBreen, my tireless, ever-patient Greek tutor. Thomas Holton kindly gave permission to use a photograph of a Rongorongo tablet taken by his late father, George Holton.

Alexander Piperski generously consented to the reproduction in these pages of his marvelous Blissymbolics puzzle, created for the 2010 International Linguistics Olympiad. Tyler Akins not only was kind enough to let us use the Dancing Men font he designed but also was gracious enough to help create a high-resolution version specifically for publication in this book.

In Britain, thanks are due to Olga Krzyszkowska and Mike Edwards of the Institute of Classical Studies at the University of London; Hannah Kendall and Amy Taylor of the Ashmolean Museum in Oxford; and the filmmaker Martin Pickles, who has long been interested in the Linear B story. Michael Ventris's biographer, Andrew Robinson, has been unstintingly generous in sharing information with me. (Though I would have dearly loved to have spoken with Ventris's daughter, Tessa, for this book, she did not respond to requests for interviews.)

Mark Aronoff, Tom Palaima, and Andrew Robinson all read the manuscript and made invaluable suggestions and corrections.

At the *New York Times,* my editors, William McDonald, Jack Kadden, and Peter Keepnews, and my Obit comrades, Dennis Hevesi, Doug Martin, Paul Vitello, and Bruce Weber, make it a joy and a privilege to come to work each day. Baden Copeland of the *Times*'s graphics department provided valuable assistance, and the *Times*'s graphics artist Jonathan Corum created this book's original maps.

Two remarkable women deserve more credit than I can ever adequately express. The first, my steadfast agent, Katinka Matson, is to be commended for the creativity and diplomacy with which she handled this book from start to finish. And my wonderful editor, Hilary Redmon, who like me has known and loved the Linear B story since girlhood, has been the only conceivable steward for this project from the very beginning to the very end. Hilary's assistant at Ecco, the supremely capable Shanna Milkey, deserves a medal for repeatedly talking me down from the ledge whenever the words "We need this image done over in high resolution" seem destined to send me fleeing there.

Suet Yee Chong, who is responsible for the elegant design of this book, juggled its welter of strange fonts with the skill of a decipherer, as did the production editor, Dale Rohrbaugh. Tom Pitoniak provided the masterful copyediting, and Nancy Wolff prepared the index.

Last but far from least, my deepest thanks go to the writer and critic George Robinson, to whom this book is dedicated, my boon companion these twenty-five years and more.

PICTURE CREDITS

Pages 1, 66, 146: Maps by Jonathan Corum. Pages 6, 39: John Chadwick, *The Mycenaean World*; reprinted with the permission of Cambridge University Press. Pages 11, 76: Ashmolean Museum, University of Oxford. Pages 16, 25: Arthur J. Evans, *Scripta Minoa*. Pages 35, 37, 70, 82, 184, 235: Evans, *The Palace of Minos*, Volume IV. Page 55, top: Digital image © The Museum of Modern Art/ Licensed by SCALA/Art Resource, NY. Page 55, bottom: Photograph by Linda Schele, © David Schele, courtesy Foundation for the Advancement of Mesoamerican Studies, Inc., www.famsi.org. Page 56: Photograph by George Holton. Page 61: Thomas G. Palaima; photograph by Beth Chichester. Page 83: Brooklyn Public Library, Brooklyn Collection. Pages 99, 101–2, 230, 317: Courtesy Alexander Piperski. Pages 105, 171, 178: Program in Aegean Scripts and Prehistory, University of Texas, Austin; photographs by Beth Chichester. Page 109: Program in Aegean Scripts and Prehistory, Department of Classics, University of Texas, Austin; photographs by Beth Chichester. Page 121: Alice E. Kober, "Evidence of Inflection in the 'Chariot' Tablets from Knossos" (1945), Figure 1; courtesy *American Journal of Archaeology*. Page 201: Photograph by Tom Blau, Camera Press London. Pages 226, 237: Institute of Classical Studies, School of Advanced Study, University of London. Page 253: Michael Ventris, "King Nestor's Four-Handled Cups," *Archaeology* 7:1 (1954); used by permission.

INDEX

Page numbers in *italics* refer to illustrations.